Springer Complexity

Springer Complexity is an interdisciplinary program publishing the best research and academic-level teaching on both fundamental and applied aspects of complex systems—cutting across all traditional disciplines of the natural and life sciences, engineering, economics, medicine, neuroscience, social and computer science.

Complex Systems are systems that comprise many interacting parts with the ability to generate a new quality of macroscopic collective behavior the manifestations of which are the spontaneous formation of distinctive temporal, spatial or functional structures. Models of such systems can be successfully mapped onto quite diverse "real-life" situations like the climate, the coherent emission of light from lasers, chemical reaction-diffusion systems, biological cellular networks, the dynamics of stock markets and of the internet, earthquake statistics and prediction, freeway traffic, the human brain, or the formation of opinions in social systems, to name just some of the popular applications.

Although their scope and methodologies overlap somewhat, one can distinguish the following main concepts and tools: self-organization, nonlinear dynamics, synergetics, turbulence, dynamical systems, catastrophes, instabilities, stochastic processes, chaos, graphs and networks, cellular automata, adaptive systems, genetic algorithms and computational intelligence.

The three major book publication platforms of the Springer Complexity program are the monograph series "Understanding Complex Systems" focusing on the various applications of complexity, the "Springer Series in Synergetics", which is devoted to the quantitative theoretical and methodological foundations, and the "Springer Briefs in Complexity" which are concise and topical working reports, case studies, surveys, essays and lecture notes of relevance to the field. In addition to the books in these two core series, the program also incorporates individual titles ranging from textbooks to major reference works.

Indexed by SCOPUS, INSPEC, zbMATH, SCImago.

Understanding Complex Systems

Pietro Auconi · Guido Caldarelli · Antonella Polimeni

Systems Orthodontics

From Clinical Reasoning to Computation, and Back

Springer

Pietro Auconi
Rome, Italy

Antonella Polimeni
Sapienza University of Rome
Rome, Italy

Guido Caldarelli
Ca' Foscari University of Venice
Venice, Italy

Institute of Complex Systems (ISC)
Consiglio Nazionale delle Ricerche (CNR)
Rome, Italy

ISSN 1860-0832 ISSN 1860-0840 (electronic)
Understanding Complex Systems
ISBN 978-3-031-64931-8 ISBN 978-3-031-64932-5 (eBook)
https://doi.org/10.1007/978-3-031-64932-5

This Springer imprint is published by the registered company Springer Nature Switzerland AG
The registered company address is: Gewerbestrasse 11, 6330 Cham, Switzerland

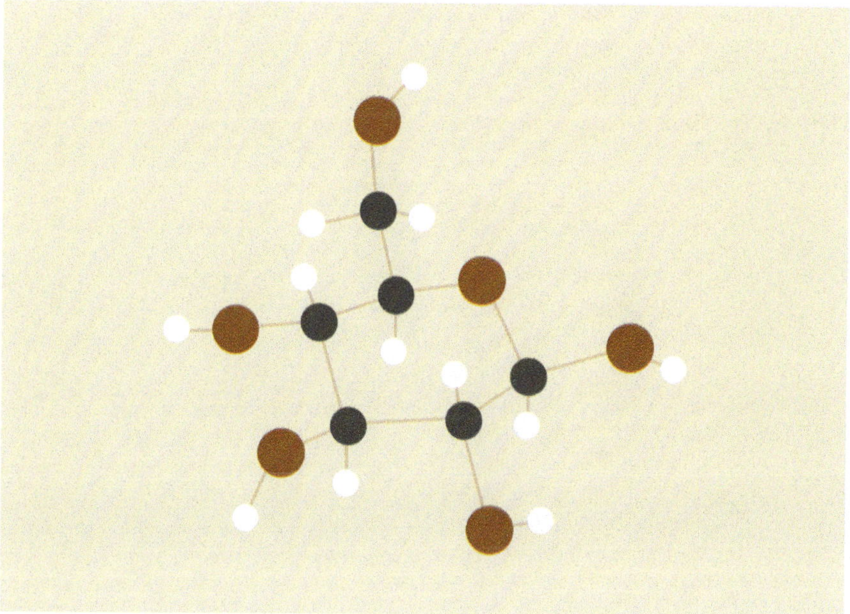

Contents

1 Introduction 1
1.1 Orthofield 4
1.2 Malocclusions 5
1.3 Self-Organisation 6
1.4 Systems Thinking and Complexity 7
1.5 Signals and Noise: The Use of AI 12
References 14

2 Cephalometric and Complexity 17
2.1 Systems Thinking 18
2.2 Understanding Facial Development and Malocclusion 22
2.3 Craniofacials Complexity 24
2.4 The Craniofacial Apparatus 28
2.5 Craniofacial Complex Evolution 30
2.6 Craniofacial Networks 40
References 45

3 Seeing the Invisible 47
3.1 The Science of Combined Effects 56
3.2 Make Cephalometric Quantities Mean More 63
3.3 Accumulation of Causes 69
References 87

4 Modeling the Invisible 93
4.1 Principles of Organised Complexity in Orthodontics 93
4.2 Tangible and Intangible Structures 99
References 117

5 Phsycal Constraints 119
5.1 More Lessons from Complexity 120
5.2 Noise and Reaction in Orthodontics 121
5.3 The Complementary Nature of Malocclusion 127
5.4 A Matter of Trust 134
References 139

6 Orthodontic Digital Phenotype: Make Data Mean More 143
6.1 The Problematic Interface Between Machine Learning and Orthodontics 145
6.2 Personalised diagnosis 149
6.3 Networks Orthodontics 150
6.4 Search for Organising Principles 155
6.5 Action Does Not Always Equal Reaction 159
References 160

Chapter 1
Introduction

All parts of the organism form a circle.
Therefore, each part is both the beginning and the end.
Hyppocrates (460-370 BC).
"Knowing how an adjustment can be achieved is useful.
Not knowing how an adjustment can be achieved is harmful.
灵枢经 Lingshu Jing (The spiritual pivot) (2600 BC ?) [1]

The burgeoning field of orthodontics presents clinicians with a deluge of new knowledge and treatment options. This "information explosion," often stemming from outside sources, combined with an ever-growing arsenal of tools and patients' diverse motivations, creates a complex decision-making landscape for orthodontists. Human cognition has limitations, typically handling only seven variables at once. Orthodontic cases, however, can involve many more. Appearances can be deceiving: the sheer number of variables and their interplay over time render averages and general rules unreliable. Orthodontic challenges often present qualitatively, meaning we cannot find normalcy by comparison but within the individual.

Furthermore, information quality could be better, with subtle, non-linear processes adding to the confusion. Imperfect observation further complicates matters. A minor initial error in landmark placement can snowball during growth, leading to inaccurate predictions and potentially disastrous outcomes. Additionally, around 20% of children lose their initial facial "memory" as development progresses, as chaotic mechanisms override the skeletal system's regularity.

Orthodontists navigate a dynamic, intricate, and sometimes unpredictable landscape. While knowledge and tools abound, careful consideration and individual-centred approaches are crucial for navigating this complexity and achieving optimal outcomes [2]. Based solely on clinical and radiographic data, the orthodontist has no way of early identification of patients changing growth trajectories [3–7], We stand at a pivotal moment in our understanding of ourselves and the world. Our reality is no longer one of the simple narratives but a complex ecosystem of interconnected variables, where each element ripples through the whole. Unforeseen events like global pandemics shake our confidence in our ability to predict and control. Compared to past eras, with their limited possibilities, our world presents a dizzying array of variables, rendering traditional risk assessment models obsolete.

P. Auconi et al., *Systems Orthodontics*, Understanding Complex Systems,
https://doi.org/10.1007/978-3-031-64932-5_1

Faced with this elusive and interconnected reality, many turn to complexity theory and its powerful partner, artificial intelligence, for answers. This "man-machine match" holds immense potential, illuminating the hidden connections between diverse phenomena across various disciplines. It unveils a tapestry of specificity, weaving together seemingly disparate threads into a coherent narrative.

When we apply computational methods in the medical field, they strive to identify recurring patterns in the past behaviour of patients. Such patterns help make predictions. These machines enable doctors to access aspects of reality more profoundly and comprehensively than simple clinical observation. But only with conditions. Future health forecasting based on patients' past data assumes that the past determines what happens in the future: we shall soon see that this assumption presents several problematic aspects once applied to the orthodontic discipline. Predicting future dental problems requires two key ingredients: a detailed understanding of the past and insightful rules to analyse it.

Firstly, predictive algorithms need access to comprehensive data on past dentofacial structures and functions, like digital scans and growth records. This data serves as a historical reference point.

Secondly, we want usable knowledge arising from this data. In orthodontics, this involves grouping patients with similar facial features and growth patterns, creating a "family tree" for comparison.

Orthodontists can identify potential future issues by comparing a specific patient's morphology to these groups. Traditionally, three main categories of malocclusions ("archetypes") served as this comparison point. However, with modern technology, we can delve deeper. Algorithms deal with big data, uncovering hidden connections between facial features, medical history, and potential future problems. This fact opens the door to individualised predictions of malocclusions even for children in their mixed or deciduous dentition, answering a question often posed by anxious parents [8–13]. Complexity theory [14] (related to "chaos theory" [15], and "theory of dynamic systems" [16]) describes seemingly unconnected systems whose behaviour does not depend upon the behaviour of a single constituent element. Imagine a bustling anthill, a vibrant city, or even your own body made up of billions of cells. These are all examples of complex systems where many individual parts interact in intricate ways.

While these systems may seem chaotic initially, an outstanding order emerges from these countless interactions. Think of it like a swarm of bees working together, each small but collectively achieving astonishing feats. This "swarm intelligence" is a hallmark of complex systems. For example, an anthill needs a certain number of inhabitants and interactions to assess its needs, like finding food. Similarly, plants compete for sunlight, exhibiting complex, collective behaviour. Interestingly, even the seemingly unrelated process of teeth exchange in our mouths shares similar dynamics to these larger systems. Dental elements "compete" for space, creating a complex interplay that shapes our smiles.

To apply the logic of complex systems to the facial system, orthodontists do not have to know every intricate characteristic of the constituent elements to understand how the dentofacial system works and progress [17] they must refer to the general, systemic laws underlying the auto-organisation of living systems, to hidden competencies focused on maintaining homeostasis, and to spatial and temporal regularities mixed with indeterminacy and random facial growth fluctuations. Complexity theory helps to understand how each piece of information behaves about its surroundings in a system where each element is strictly dependent on the others. The effects originating from the mutual interaction between the elements generate "emergent properties" and local organisational zones that behave as *basins of attraction* and canalisation of defects ("patterns of contagion"). A grand unified theory of complex systems still needs to be included; after all, it is not easy to find a formula to describe and formalise the movements that occur in an ant colony, in organising a tornado, or in the transactions of traders on Wall Street. Reading these phenomena requires a decentralised model of thinking, which sees the development of these systems as the effect of an interactive phenomenon between local organising agents and global constraints. During facial growth, small changes in provincial zones and growth anticipations or time delays in critical areas can produce vastly different final skeletal scenarios. This book explores these phenomena, conducted with unprecedented investigative skills offered by computer science.

Orthodontics is the science of change over time and of consistency of the facial system concerning change. The orthodontist professes an inherent holistic science that includes balance, harmony, mutual protection between components, stability, acceptable developmental compromises, and more. They must find the best possible combination of these components before making decisions. They seek linear explanations to more complex problems and attempt to define general rules, reproducibility of phenomena, and alignments of causes and effects. But more is needed.

Intervening at the right time using the appropriate corrective tools on the proper facial areas makes it possible to reorient the existing facial pattern, guide the adaptive remodelling of the facial components, facilitate the resolution of local discrepancies, release the proprioceptive stimuli necessary to achieve the natural organisation and propensity of the masticatory organ: the best occlusal-skeletal equilibrium.

Complexity theory and computer tools procedures add evidence to the equilibrium concept. Dentoskeletal equilibrium is the state of stillness related to the resultant of the forces acting on the components of the system and the forces external to the system. The quantitative measurement of these forces could be more problematic in static and dynamic situations, as are controlled experiments to clarify the disparate aspects of tooth movements under the influence of those forces [18]. The balance results from the intraoral forces acting on the crowns and the forces acting on the root from the surrounding bone through the periodontal ligament. The troops (muscular, occlusal, adjacent tooth elements, etc.) must be in equilibrium to maintain the system's stability over time. But this is only sometimes the case. Sometimes, there is no clear relationship between the forces we apply and tooth movements. We know less about the subject than we think we do.

1.1 Orthofield

In this book (Chap. 3), we will see that defining a concept of optimised dental equilibrium in the form of overall system harmony is not trivial, despite the help of computational algorithms, since we need to define a basic idea as a relational form of the system. And yet, that is what the orthodontist wants. Orthodontists must deal with a "Orthofield" surrounding the dentofacial complex, acting on the elements of that complex in a manner analogous to the action of a physical field [19]. As a magnet is surrounded by a force field influencing everything within its range, the "orthofield" in orthodontics can be considered a dynamic zone of hidden forces constantly shaping teeth and jaws as they grow.

We must consider how this "orthofield" evolves in space and time during growth for successful orthodontic treatment. Timing is crucial. Interrupting these forces at the wrong moment can cause the original growth pattern to reassert itself, potentially hindering desired outcomes.

The morphogenetic facial pattern contains potentials that are not immediately clinically manifest because there is a stop to their activity. Over time, different phenomena of intrusions in the system, such as oral respiration or adverse habits, can disturb or delay the appearance of the morphogenetic pattern. Moreover, we cannot take for granted that the patient's orthofield agrees with our treatment plan, nor that the timing of the unfolding agrees with our interceptive attitude to dysmorphisms. We try to establish harmony, balance, and optimal aesthetic relationship, but we need to know in depth what the proper balance expressed by the orthofield is. We need to know all the ingredients of these abstractions. An untimely interceptive intervention may be performed before a distortive orthofield has fully emerged, making relapse inevitable (never consider the facial aesthetics of the child as the aesthetics of the adult) [19].

When studying the face, we deal with complex concepts like how features form, how they vary among people, and how they interact. This feature creates a gap between the "average" measurements we often rely on and the unique reality of each patient. Despite this limitation, we can still predict future growth based on their facial structure. However, this prediction isn't a fixed path like a train schedule but rather a map highlighting various possibilities. We select specific features and differences to analyse, transforming them into valuable information. However, to truly understand how individual genes translate into specific facial features (genotype to phenotype), we need better tools to explore the intricate processes involved.

Complexity is endemic in orthodontics, so for each phenomenon we see, we must expect an opposite phenomenon and an opposite statement, which is not wrong but only has less plausibility. Our patient's representation is inevitably idealised and seasoned with a large margin of inaccuracy. We efficiently run into many sources of bias: incorrectly identified patients, missing data, insufficient samples, underestimates, and misclassifications. We do not know what influence the altered parameters have on other parameters or the preferential corridors for spreading defects. The degree of knowledge we have about the craniofacial system is related to what we know about the information shared between system components. We consider the facial system an integrated and stable unit with a solid *morphological integration* that does not always protect against disharmony.

No single relief can fully describe a complex system: one to two degrees or millimetres of error are generally allowed for each angular or segmental cephalometric measurement so that even the best knowledge of the patient is inevitably approximate knowledge. But how wrong is the approximation? The internal structural organisation of the craniofacial system, together with stimuli from the environment, random growth events, and adaptive capacity, channels global facial changes through specific directions. As stated, models based on the laws of biomechanics explain only some things. The balanced model we impose on the patient is an arbitrary arrangement which the patient's orthofield may not entirely accept. Currently, research and practising orthodontists can transfer these processes into the computable realm, using intricate mathematical manipulations leading to continuously updated descriptive and predictive probabilistic models [20]. The appreciation of the whole ortho field drives us to pay more attention to the individual patient and to adopt methods of treatment which emphasise personal response to the general laws of orthodontics.

If we want to define who is abnormal from an orthodontic point of view, it is necessary to have a clear idea of normality. The laws of statistics are amongst the best available methods for scientific research. The reference to average angles of inclination of the *dental axes, overjet, overbite, and proximal horizontal & vertical contacts* between tooth components should be considered suitable for every patient. Some patients may exhibit a discrepancy in the sum of diameters of the upper and lower incisors, which would indicate a variation in the degree of acceptable or preferable tooth inclination, overbite, and overjet [19]. There may be a discrepancy between the basal bone and the teeth, which may recommend extractions in one patient and contraindicate them in another. Moreover, the different verticality of the skeletal pattern imposes different optimal inclinations of the dental elements.

Facial morphology throughout life is so unsettled that a stable representation is problematic. All system components are variable, and the only definable thing that remains stable over time is the shape of the teeth. The orthodontist pursues the ideal dental occlusion, hoping to achieve a better balance of all surrounding structures, putting the muscles in the best possible contact with the proprioceptive reflexes generated by the dental cusps. These criteria will fit as many patients as possible. Yet, relapses are frequent annoying problems for the orthodontist. One more time, cephalometry doesn't tell us the whole story.

1.2 Malocclusions

Malocclusion expresses a peculiar form of equilibrium, a bottleneck in the defect disposal flow of the craniofacial system. It contains hidden information about the operating morphological generative forces we must fight against. Network science offers the possibility of mapping the topology of facial morpho-functional dynamics and defining the direction of competitive or collaborative influences between different skeletal zones. Currently, in the orthodontic discipline, a change in the system of thinking is taking place about the choice of treatment to be used to guide atypical facial growth back to normal. We have just said that the operator cannot refer to stable, perennial, and reproducible orthodontic morphologies, which can fix patients into defined categories linked to prescribed forms and progressions. And that the division into Class I, II, and III malocclusions represents a simplification of a more complex reality. In complex biological systems, form couples with the process; opposing components cooperate through morphological oscillations during the growth process, and collisions of morphologies and temporal fluctuations of constituents represent 'memory mechanisms'. Thus, it takes work to get a reproducible picture of the system you are dealing with.

Wherever processes of growth and morphological configuration of living matter take place in nature, information must be present since every morphology is the formal expression of a content: "...form, structure and movement of any entity are merely a materialisation of energy, more or less concentrated and in transformation" [21]. This energy leaves traces in each facial morphology: in biomechanical information, force, mass, shape, and energy are not separable, as information transmission needs a material medium. We will see why the global structure of the dentoskeletal system couples to local morphological characteristics and why we must consider local data about the characteristics of the whole skeletal imbalance.

1.3 Self-Organisation

The *self-organisation* of complex systems gives rise to the spontaneous formation of space-time feature *modules* (also *motifs,* or *communities,* in Network analysis, see Fig. 3.3), groups of densely associated components connected with loose links related to antagonism or cooperativity among components. In biological systems, modules have always been considered ubiquitous, elementary forms of organisation. We will see their role during facial growth. The self-organisation in complex systems represents how simple rules determine complex behaviours. In nature, it happens when, in different dimensions, the cells of a growing organism aggregate to form tissues and organs. The desert wind produces circular sand structures, as simple local rules [22], with the participation of gravity, wind, and friction between grains of sand [23]. But the simplicity of rules can be illusory; it might give us the impression that we can derive simple explanatory stories for complicated phenomena.

To make sense of the things of the past, we should be able to separate the past stable trend from past random deviations. The interaction between dentoskeletal parts is determined by growth and collisions between structural differences in the facial system, according to reciprocal chains of growth-form determination. Describing these phenomena requires a reliable classification of morphological and relational entities to be offered to computational systems. The higher the uncertainty of the process, the more noise exists in past information that needs to be ignored or probed somehow in the ways we see. In Chap. 4, simple algorithmic procedures such as *Fast and Frugal Trees* can improve response accuracy about growth and treatment outcomes by minimising irrelevant or misleading information. This procedure allows small patient samples and less processing to produce better inferences [24]. The goal is to determine the environmental structures in which a given treatment decision shortcut (*heuristic*) is booming despite the superficiality of the analysis.

In daily professional life, orthodontists do not solve the patient's problem once and for all, they have to solve it continuously over time. This approach is essential because orthodontic issues tend to recur, and the face of the child is constantly changing. Some problems are simplistically read as side effects: in reality, they were problems and processes that had not been visualised and which later come to the attention of the clinician. The bite opening may not be due to the treatment but to oral respiration or incorrect tongue posture. At each evolutionary stage, new laws, concepts, and generalisations are needed.

Heuristics cause the programme to focus on only part of the search space. These forms of mathematical simplification run the risk of representing probabilities in the data in a simpler way than they are. In orthodontics, the i.i.d. hypothesis commonly used by analysts (the probabilities in the data are independent and identically distributed) may be an oversimplfication due to the presence of very different morphological constraints between patients.

The order of living matter is an order of morphological forms; where there is a form, one moves away from homogeneity, disorder, and entropy. Codes linked to the patterns of space-time structures contain computational information: once received, the information allows decisions between different possible evolutionary choices ("evolutionary bifurcations"). Some dominant defects propagate their information to nearby elements over time, following contagion patterns resembling those of infectious diseases. Such a similarity allows doctors to hypothesise the contours of the "possible adjacent", the most probable morphological evolution and canalisation of local skeletal components. Some elements in the network are compartmentalised, insensitive to the stimuli of the neighbours, and therefore tend to remain "invariant". Every defect found in the system plays the role of the centre of convergence of a series of past or current, actual or possible, probable or improbable relationships. Cephalograms provide a geometric framework of the existing skeletal structure as a sum of past development, but geometric data alone does not provide predictive information. Yet, as we will see in the next chapter, some predictive orientation is possible by following the network of influences and the cumulative effect of the skeletal arrangement on growth [25, 26].

1.4 Systems Thinking and Complexity

If in a complex system, "the whole is more than the sum of the parts", the meaning of a skeletal characteristic is perceived in a broader context, to a more comprehensive function, or a system's whole facial *propensity*. The state of a growing complex system depends on the previous one and the conditions of the following one. Some features propagate through specific pathways that work as 'sources' of defects, disposed of in different 'sink' structures. *Systems thinking* starts with the assumption that it is not so much the single abnormal variable that characterises the outcome but rather the interconnection between structures. Although we cannot predict all the final endoskeletal details, the general shape of the new morphologies and the new forms of organisation may be relatively predictable and straightforward.

In biomedical research, more than simply measuring things is required. We must delve deeper into what those measurements tell us about how things work and why. Network analysis is a robust framework for that purpose. As applied to orthodontics, it shows us how everything in the face and jaw connects—directly or indirectly, strongly, or weakly. Think of it like a complex web where everything somehow influences everything else. Complexity theory takes this a step further. It studies how patterns and structures emerge from this web of interactions. So, for orthodontists, the key is to:

1. **Identify the essential pieces**: What parts of the face and jaw are we studying?
2. **Look for patterns**: How are these parts connected and interacting at a specific time?
3. **Conclude**: What underlying principles govern these interactions?

By doing this, we can move beyond just measurements and better understand how the face and jaw develop and change. The contingency is ineradicable as living matter exhibits a combination of simple fundamental laws combined with the operation of chance. The logic of contingency teaches us not to limit ourselves to understanding what can happen considering what has happened but also to consider the counterfactuals, the possibilities lost in that event (patient), which instead found expression in another [27].

New approaches in science are helping us understand complex systems like the human body. Using math from "complexity science," we can combine complex data with individual details to create a more complete picture of a patient's health. This procedure allows us to see beyond apparent symptoms and understand the hidden connections between different body parts. This idea is beneficial when predicting how someone might develop, particularly when we must fully understand the underlying processes. However, it is essential to remember that computers cannot replace human judgment. While algorithms can analyse data and identify patterns, they need more biological understanding and foresight for complex systems. They can locate "competition" between different parts of the body, but they cannot grasp the bigger picture or predict future consequences. Ultimately, orthodontic treatments aim to restore balance and harmony within the body. The treatments can sometimes create temporary disruption, but the goal is to minimise disharmony and encourage healthy development.

As Sheldon Baumrind said, "Orthodontic therapy is applied biology". Indeed, the common denominator of the onset of cytoskeletal disharmony is that it is a collective biological process born from repeated local interactions that generate collective information and mutual cascading effects on the entire system. The factors that persist the longest significantly impact the system [28]. As the dysmorphosis spreads, the whole system enters a new *phase:* the challenge is to anticipate the transition process between acceptable and unacceptable facial morphology, choosing the right tools and procedures to identify the underlying latent phenomena before they cause an overall distorting effect. In any case, the ultimate purpose of the system's whole proprioceptive information is to safeguard and optimise occlusal contact.

We must rephrase that orthodontists must look for a rule that values past craniofacial observations in predicting future events. They must look for historical data vectors applicable to the problem space. Since orthodontics is a discipline of contingencies rather than general rules, nothing can make an orthodontic treatment based on misleading contingencies assumptions efficient. Framing contingencies and giving them proper weight is the first step. Computational machines extract patterns from data regularities; in orthodontics, randomness can overwhelm regularity. In a dynamic system subjected to flows of biomechanical forces, relationships of domination, or mutual congruence between variables, are formed. A deep Spee curve determines an extrusion of the upper premolars as much as the extrusion determines a deep Spee curve. The relationships of forms are associated with relationships of forces. Geometric *attractors* respond to the topology of relationships: some are subject to the forces of growth, others simply to forces generated by the meeting of reciprocal physiognomy.

Rather than underlying processes, orthodontic "laws" are expressions of constraints acting on developmental individual potential. An overjet can have an additional dimension in an insufficiency of the vertical dimension. An incisal protrusion can have a further dimension in camouflaged natural crowding. A large Goniac angle indicates the mandibular ramus's difficulty in keeping up with the vertical growth of the alveolar district. In the ontogenetic skeletal model, it is necessary to include the formal connections between current morphologies and past events, specifying the dynamics that led to what is visible in the system. Some computational models may deconstruct the facial system, others a re-composition of its constituents and past events.

As stated, many aspects of orthodontic reality present themselves to our gaze in networks, i.e. aggregates of nodes connected in a dense manner (See Fig. 4.5). This state of things allows a conceptual framework favouring the minimisation of contingent growth phenomena. As the interdependence structure between variables can represent craniofacial morphology, the *topology*, i.e., the structural organisation, constrains the range of possible behaviours and describes the information in different system zones. Due to the need for more sufficiently general formalism in craniofacial research, understanding structure-dynamics relationships is still a significant goal. The Network analysis can be a response to understanding how seemingly insignificant variables can become crucial once combined with other variables.

In every situation of uncertainty, probabilities are involved: the advantage of probabilistic reasoning over logical, analytical rationale lies in the possibility of arriving at rational descriptions, even when there is not enough deterministic information about how the system works. If two events, x and y, are highly correlated, it is reasonable to assume that knowing that x is accurate can help understand the probability of y. Network analysis discovers the most influential probabilities among the variables, i.e., the probable coevolution, and can remove the least influential ones that do not reach the acceptable correlation threshold. But the actual data to offer the computational system should be the data conjugated with its background noise, skeletal constraints, residual growth, reciprocal collisions of forms, lean zones each component influences or is influenced by, and so on.

Change is an intrinsic quality of living matter. All parts of a biological system have a turnover: in the human body, the pancreas replaces its cells every 24 h. What counts is the state of the pancreatic system, which must maintain its functions over time, not individual cells. Living systems perform a continuous optimisation process of improvement of adaptive capacities. Computational systems express a tension towards optimisation of their level of competence through a constant adjustment of the assimilation process of the elements. We will consider offering Artificial Neural Networks cephalometric data to obtain information on the need for problematic circumstances, such as dental extractions. The data learned in each layer of neural network computation passes to subsequent layers towards increasingly complicated levels of abstraction. In doing so, the learning machine modifies its internal state by changing the strength of certain connections. It looks for regularities in the data to extract patterns. However, as stated, the realm of orthodontics is one of irregularities. Thus, there is a risk that, instead of being an operational decision support, excessive multiple abstractions could paradoxically distance the orthodontist from the empirical reality of the patient, cast in a cloud of different combinations of signs, symptoms, evolutionary paths, conditional dependencies, and adaptive processes. When it comes to excluding any quality that is not numerical and considering patients as pure quantitative multiplicity, there is the risk of identifying as consistent patterns simple random feature fluctuations that do not represent any underlying structure [29].

As in any complex biological system, the facial system can exhibit an *emergent* behaviour that does not derive from the morphological characteristics of specific constituents. It emerges naturally from the organising relations of the parts. This phenomenon forces the orthodontist to reconsider the value of the parameters: we will see that a high degree of membership in a morphological cluster implies that the robust connectivity acting into the cluster contains more growth information than the absolute value of individual features at precise moments in time. Thus, membership is truly a biological characteristic.

Orthodontic preventive surveillance consists precisely of preventing the system from entering a *self-organised criticality* [30] during the growth process and bearing in mind that by applying a therapeutic device, we act both on the anatomy of the system and the biology of growth. Inevitably, we extensively modify the network of interactions in the craniofacial ecosystem without knowing all the players involved and having a precise idea of the long-term effects. In a system based on instability, where elements link to each other, we must continually redo the calculations during treatment. The adaptation of facial components to the environment is not only reactive, but it also expresses a form of preventive cognition drive to plan a response to future occurrences and environmental changes. These dynamics are not very different from strategies implemented in the plant kingdom to prevent the effects of seasonal changes [31].

The operator uses computerised procedures to check the deviation from the a priori expected result, i.e., to limit the reliability of the conceptual model. They formulate hypotheses a posteriori, trying to find plausible reasons to explain how one or more elements from the past induce behaviour. Suppose we want to get the most out of the data. In that case, we need to rely on accurate, descriptive parameters and threshold risk values, knowing that morphological boundaries change over time and that what is irrelevant changes according to the situation. As in many medical fields, to clarify the natural history of dysmorphic orthodontic features (not being able to count on how many patients followed over time without undergoing treatment), it is necessary to refer to the quality of the data. We will see that valuable tricks exist to focus on high-quality data, working on subspaces of data, recurring to *synthetic patients*, reasoning in terms of global characteristics, stratifying subjects in segments by age and gender, or reasoning in terms of *qualified* patients, with highly expressive capacity about the topic of interest.

Like all organic systems, the facial system tries to overcome growth imbalance by using specific forms of homeostasis, in which opposing forces try to balance each other. The orthodontic anamnesis is not a chronological narrative of independent events but an insight into a developmental process. The most challenging step for the orthodontist clinician resides in moving from the list of problems to treatment priorities. It is the moment to synthesise the patient's wishes and realistic treatment options [32]. Treatment planning involves the clinician's subjective judgment, supplemented by consideration of the most critical problems experienced by the patient. The orthodontist should tell the child's parents that there might be different aspects to consider, less visible but crucial. Detecting a morphological anomaly in a specific facial area may be intolerable for the patient. Still, it may need to demonstrate more relevance concerning the balance of the entire facial system. It may be part of the flexibility necessary for the system to absorb more relevant but less visible pathological phenomena located elsewhere or may derive from the interconnection of other data not sufficiently considered. Facial areas located near bone growth centres function as *hubs* which cause radiation of defects to other structures. Network science visualises that each characteristic has its specific interactive map with different structures; this plays a role when formulating the judgment regarding treatment priorities [33]. While addressing orthodontic misalignment in a single spatial plane may take less than a year, most orthodontic concerns encompass diverse issues impacting skeletal structure, dental and alveolar components, and functional aspects. Notably, each such issue exhibits multidimensional characteristics. In complex systems, elements acquire environmental information and adapt their relationships accordingly. Within the facial system, the sheer volume of interconnected problems and the age-related structural increase in complexity (shifting from smooth surface contact in deciduous teeth to the interlocking cusps of permanent teeth) contribute to issues becoming not merely additive or proportional but rather exponentially complicated.

1.5 Signals and Noise: The Use of AI

Powerful tools like decision trees and artificial intelligence are helping doctors map facial features to diagnoses and treatments. These tools learn from past data to predict outcomes for new patients, offering valuable insights.

However, doctors must combine this new information with their existing knowledge and experience. Understanding the "why" behind a patient's improvement or worsening goes beyond data analysis. Doctors must consider the intricate factors at play to make the best decisions. Unlike humans, computers learn in artificial universes where everything is on equal footing. This property means they can identify patterns but may need help grasping the underlying causes. Just because they know "what" happens doesn't tell them they know "why." In medicine, focusing on individual molecules only sometimes explains the entire disease.

Similarly, assuming a direct cause-and-effect relationship from correlations can be misleading for complex systems like the human body. Machine learning algorithms don't have this challenge. They find patterns and make predictions, but they need to gain the conceptual understanding a doctor possesses. Essentially, these tools are potent allies but not replacements for human judgment. The value lies in combining the strengths of both—data-driven insights and experienced intuition—to ensure the best possible patient care. Imagine powerful tools that can analyse mountains of data to predict health risks like diseases or facial abnormalities. Unlike doctors, however, these tools (algorithms) don't understand the "why" behind their predictions. They're like black boxes that crunch numbers and give answers without explaining the logic. Think of it like searching for buried treasure with a metal detector. You get beeps based on hidden metal but don't see the metal or understand its meaning. Similarly, algorithms find "critical points" in the body that suggest specific outcomes but can't explain why or how.

However, their strength lies in seeing the bigger picture. They consider everything in the body simultaneously, knowing everything is interconnected (though not equally or simultaneously). This holistic picture allows them to identify hidden patterns and predict future problems, like disease onset or facial growth issues. Instead of following specific instructions, they explore network connections within the data, assigning importance (weights) to different factors based on their influence. These "weights" tell us how much each factor contributes to the predicted outcome. The next chapter will delve deeper into specific algorithms like Factor Analysis. These tools can predict a patient's risk of developing a disease or malocclusion by considering both evident and hidden factors—something even highly skilled doctors can only do with a different level of detail and speed.

Conversely, through experience, orthodontists develop a unique understanding of facial growth, allowing them to interpret subtle clues and identify general patterns in various cases. This "pre-empirical culture" acts as a lens through which they filter and analyse new observations. Even when data is unclear, their intuition, honed by experience, helps them form hypotheses and trace potential causes. They naturally seek the simplest explanation that fits the evidence, leveraging their innate ability to connect the dots. Doctors refine their mental model of the jaw and mouth over time, allowing them to extract meaning from even the most minor details. They understand how treatments interact with natural growth, recognising the interconnectedness and complexity of the entire system. Ultimately, the success of treatment hinges on anticipating how the facial system would develop without intervention. In other words, it requires considering hidden variables and the intricate interplay between various components [34]. While machine learning excels at finding patterns in large datasets, its predictions for individual cases based on these patterns can be unreliable due to inherent uncertainty and the unique characteristics of each patient [35]. These individual variations are crucial in orthodontics, where facial development follows unique and dynamic pathways. While some morphological patterns may persist with slight variations due to the underlying laws of development, unforeseen changes or "catastrophes" can also reshape the facial structure. To effectively navigate this complexity, orthodontists rely on two fundamental principles: understanding the natural developmental boundaries of different facial areas and analysing the likelihood of future changes in each individual. Tools like the Sakey diagram can aid in visualising these factors and guiding treatment decisions that respect the unique growth potential of each patient.

As discussed above, the facial structures and tissues are not static but continuously interact and change. This ongoing process eventually produces a cohesive and defined structure. However, this "orthodontic space" is not uniform, as different components comprise diverse materials. As living systems grow, they become more organised and structured. This growth involves processing information and reducing the potential for future variations in their development. While cephalometric data provides valuable insights, this insight is just a momentary snapshot of a constantly evolving landscape influenced by a complex interplay of skeletal, dental, muscular, and other elements. Internal mechanisms and the level of interaction between these elements can create self-reinforcing patterns, limitations, and corrective processes that make predicting future outcomes based solely on past data problematic. Ultimately, the growth process determines the range of possible outcomes and future development. Orthodontic data often come from various sources and periods, making it inconsistent. However, analysing the spatial arrangement of these elements is crucial. Different parts influence each other within specific timeframes and to varying degrees. In this complex system, everything is interconnected, with each facial feature shaped by its relationship to others. Even seemingly isolated data points gain meaning when placed in this context. Any change in one element affects the entire system, altering its overall characteristics.

This book avoids the danger of being one more manual offering a foundational approach to the diagnosis and prognosis of orthodontic discipline, even if it sometimes stands on the edges of orthodoxy. It discusses the effects produced by the reciprocal influence between biological growth constraints and morphological constraints in facial structure. This relation happens when every element, directly or indirectly, meets every other system component. It simply intends to push the orthodontist to master heterogeneous concepts and a kind of tolerant, more expanded thinking, which has become a necessity of thought in many fields today. The basic idea is that there is more to the orthodontic patient than we can see and that computers can help us understand more, providing us with mathematical documentation of the different models of dentofacial self-organisation.

References

1. B. ItsHaky, Huang Di Nei Jing Ling Shu; the ancient classic on needle therapy; the complete Chinese text with an annotated English translation. Alternative, Complem. Integrat. Med. **3**(1), 1–11 (2017). https://doi.org/10.24966/ACIM-7562/100020
2. S.E. Bishara, in *Textbook of Orthodontics* (Saunders, 2006). Accessed: Oct. 07, 2023. [Online]. Available: https://www.amazon.it/Textbook-of-Orthodontics/dp/8181470729
3. A. Rajkomar, J. Dean, I. Kohane, Machine learning in medicine. N. Engl. J. Med. **380**(14), 1347–1358 (2019). https://doi.org/10.1056/NEJMra1814259
4. R. Bellazzi, B. Zupan, Predictive data mining in clinical medicine: current issues and guidelines. Int. J. Med. Inform. **77**(2), 81–97 (2008). https://doi.org/10.1016/j.ijmedinf.2006.11.006
5. A.A. Freitas, Understanding the crucial role of attribute interaction in data mining. Artif. Intell. Rev. **16**(3), 177–199 (2001). https://doi.org/10.1023/A:1011996210207
6. P. Auconi et al., The validity of machine learning procedures in orthodontics: what is still missing? J. Pers. Med. **12**(6) (2022). https://doi.org/10.3390/jpm12060957
7. S.N. Asiri et al., Applications of artificial intelligence and machine learning in orthodontics. APOS Trend. Orthodont. **10**(1), 17 (2020). https://doi.org/10.25259/APOS_117_2019
8. K. Bahaa, G. Noor, Y. Yousif, The artificial intelligence approach for diagnosis, treatment and modelling in orthodontic, in principles in contemporary orthodontics. INTECH (2011). https://doi.org/10.5772/19597
9. R.C. Deo, Machine learning in medicine: will this time be different? Circulation **142**(16), 1521–1523 (2020). https://doi.org/10.1161/CIRCULATIONAHA.120.050583
10. R. Miotto, L. Li, B.A. Kidd, J.T. Dudley, Deep patient: an unsupervised representation to predict the future of patients from the electronic health records. Sci. Rep. **6**, 26094 (2016). https://doi.org/10.1038/srep26094
11. T. Pelaccia, G. Forestier, C. Wemmert, Deconstructing the diagnostic reasoning of human versus artificial intelligence. CMAJ **191**(48), E1332–E1335 (2019). https://doi.org/10.1503/cmaj.190506
12. E.J. Topol, High-performance medicine: the convergence of human and artificial intelligence. Nat. Med. **25**(1), 44–56 (2019). https://doi.org/10.1038/s41591-018-0300-7
13. D. Weinberger, in *Everyday Chaos: Technology, Complexity, and How We're Thriving in a New World of Possibility* (Harvard Business Review Press, Boston, Massachusetts, 2019). Accessed: Jan. 03, 2023. [Online]. Available: http://search.ebscohost.com/login.aspx?direct=true&db=bth&AN=135705805&site=ehost-live&scope=site%0A; http://hbr.org/product/a/an/10168-HBK-ENG
14. P. W. Anderson, More is different. Sci. **177**(4047), 393–396 (1972). https://doi.org/10.1126/science.177.4047.393
15. E. N. Lorenz, Section of planetary sciences: the predictability of hydrodynamic flow*,†. Trans. NY Acad. Sci. **25**(4) Series {II}, 409–432 (1963). https://doi.org/10.1111/j.2164-0947.1963.tb01464.x
16. C. Grebogi, E. Ott, J.A. Yorke, Chaos, strange attractors, and fractal Basin boundaries in nonlinear dynamics. Sci. **238**(4827), 612–638 (1987). https://doi.org/10.1126/science.238.4827.632
17. T.M.O. Forum, The nature of an orthodontic diagnosis. Am. J. Orthod. **50**(7), 490–503 (1964). https://doi.org/10.1016/0002-9416(64)90156-3
18. S. Weinstein, et al.: On an equilibrium theory of tooth position. Angle Orthodont. **33**, 1–26 (1963)

19. H. Schwartz, The field concept in orthodontics. Angle Orthodont. **30**(3), 154–161 (1960)
20. K.I. Goh, M.E. Cusick, D. Valle, B. Childs, M. Vidal, A.L. Barabási, The human disease network. Proc. Natl. Acad. Sci. USA **104**(21), 8685–8690 (2007). https://doi.org/10.1073/pnas.0701361104
21. J. Rifkin, T. Howard, N. Georgescu-Roegen, in *Entropy: Into the Greenhouse World* (Bantam Books, 1989). Accessed: Oct. 06, 2023. [Online]. Available: https://www.amazon.it/Entropy-Greenhouse-World-Jeremy-Rifkin/dp/0553347179
22. M. Forster, E. Sober, How to tell when simpler, more unified, or less ad hoc theories will provide more accurate predictions. British J. Philosophy Sci. **45**(1), 1–35 (1994). https://doi.org/10.1093/bjps/45.1.1
23. L. Fisher, in *The Perfect Swarm: The Science of Complexity in Everyday Life: Fisher* (Len: Amazon.com.be: Books. Basic Books, 2009). Accessed: Oct. 06, 2023. [Online]. Available: https://www.amazon.com.be/-/en/Len-Fisher/dp/046501884X
24. G. Gigerenzer, H. Brighton, Homo heuristicus: why biased minds make better inferences. Top. Cogn. Sci. **1**(1), 107–143 (2009). https://doi.org/10.1111/j.1756-8765.2008.01006.x
25. S. Baumrind, Current clinical research in orthodontics: a perspective. J. Calif. Dent. Assoc. **34**(10), 831–834 (2006). https://doi.org/10.1080/19424396.2006.12222254
26. P. S. Vig, in *Science and Clinical Judgment in Orthodontics* (1986). Accessed: Oct. 06, 2023. [Online]. Available: https://deepblue.lib.umich.edu/handle/2027.42/146712
27. C. Marletto, in *The Science of Can and Can't: A Physicist's Journey Through the Land of Counterfactuals* (Penguin Books, 2021). Accessed: Oct. 06, 2023. [Online]. Available: https://www.hoepli.it/libro/the-science-of-can-and-can-t/9780141986463.html
28. J.C. Ramo, in The *Age of the Unthinkable: Why the New World Disorder Constantly Surprises Us And What We Can Do About It: Ramo* (Joshua Cooper: 9780316118088: Amazon.com: Books. Little Brown and Company, 2009). Accessed: Oct. 06, 2023. [Online]. Available: https://www.amazon.com/Age-Unthinkable-Disorder-Constantly-Surprises/dp/0316118087
29. F. Cabitza, in *Breeding Electric Zebras in the Fields of Medicine* (2017). https://doi.org/10.48550/arXiv.1701.04077
30. P. Bak, C. Tang, K. Wiesenfeld, Self-organized criticality: an explanation of the 1/f noise. Phys. Rev. Lett. **59**(4), 381–384 (1987). https://doi.org/10.1103/PhysRevLett.59.381
31. P. Calvo, N. Lawrence, in *Planta Sapiens* (2022). Accessed: Oct. 06, 2023. [Online]. Available: https://www.ilsaggiatore.com/libro/planta-sapiens
32. W.R. Proffit, in *Contemporary Orthodontics* (Mosby-Year Book, 1993). Accessed: Oct. 06, 2023. [Online]. Available: https://www.worldcat.org/title/contemporary-orthodontics/oclc/489919510
33. A. Scala, P. Auconi, M. Scazzocchio, G. Caldarelli, J. McNamara, L. Franchi, Complex networks for data-driven medicine: the case of class III dentoskeletal disharmony. New J. Phys. **16**(11), 115017 (2014). https://doi.org/10.1088/1367-2630/16/11/115017
34. P.H. Buschang, R. Tanguay, A. Demirjian, L. LaPalme, J. Turkewicz, Mathematical models of longitudinal mandibular growth for children with normal and untreated Class II, division 1 malocclusion. The European J. Orthodont. **10**(3), 227–234 (1988). https://doi.org/10.1093/ejo/10.3.227
35. M. Blastland, in *The Hidden Half: The Unseen Forces that Influence Everything* (Atlantic Books, 2020)

Chapter 2
Cephalometric and Complexity

Instead of decreasing the dimensionality of the dataset by removing less significant data, it is possible to use *data augmentation* procedures, which extract virtual patients from actual orthodontic patients, to improve the predictive ability of the algorithms. Each actual patient can be matched with similar synthetic patients, becoming its cluster [1]. The natural malocclusion history requires subsequent measurements in the same patient. Given the difficulty of having patients followed over time without undergoing therapy, one can understand the preciousness of patient simulation and augmentation methods. Focusing solely on individual details in an overly analytical way can lead to missing the bigger picture. In living systems, self-organisation is analogous to understanding a "symptom pattern." Similar to how disease symptoms are not just a collection of signs but reflect a more profound, unified response by the organism, facial development also involves interconnected elements dynamically changing.

The orthodontist's role is to create a computational model that captures this new dynamic unit, incorporating the unique features of the specific situation. This active unit influences and constrains the individual parts of the facial system and their growth patterns. During the growth process, vertical measures of dentofacial complex usually show more significant random variance than horizontal measures. The orthodontist must define the feature stability and how much an abnormal local feature can modify the global dynamic structure. It can become a "conducting part", which generates and propagates "states of tension", or it can be reabsorbed by rebalancing behaviours on the part of other structures.

The two facial biotypes of Class II malocclusion divisions 1 and 2 are different, as are physiology, skeletal development, and prognosis. Often, the biotype of Class II malocclusion, division 1, is associated with a dolichofacial or mesofacial morphology, labial incompetence, oral respiration, narrow jaw, retrognathic jaw, incorrect lingual posture, increased facial convexity, short mandibular ramus, vertical occlusal plane, mandibular post-rotation. In contrast, the biotype of division 2 is meso- or brachyfacial, associated with normal mandibular ramus, deep bite,

P. Auconi et al., *Systems Orthodontics*, Understanding Complex Systems,
https://doi.org/10.1007/978-3-031-64932-5_2

muscular hypertonia, and average convexity. These are two distinctly different global structures, united only by occlusal relationships. Thus, the models for analysing what happens in the first and second divisions of this malocclusion must be other, with different evolution and functional qualities, in which the numerical value of each characteristic will depend on the overall facial characteristics. The possible feature combinations expressed in the craniofacial biotype limit a conceptual synthetic definition suitable for describing each patient's behaviour. The matter is further complicated because, as an example, a dolichofacial patient with Class II malocclusion 1 division has little chance of catching up on growth: his mandibular growth potential is poor, often there is respiratory obstruction, atypical swallowing, allergy, vicious habits and often, even worse, the direction of growth is vertical. This information can be crucial in visualising the outcome of development and treatment success. Knowing how to read the contribution of each of them and, even more, their interaction on the overall facial morphology offers attractive possibilities.

2.1 Systems Thinking

As complex systems exhibit emergent qualities absent in any of the parts, *systems thinking* shifts the focus from the parts to the whole. A system is a set of interacting features that generate something more significant than the sum of the components together. Everything is a system in real life. Within each problem, there is always a bigger picture, a pattern, and a recurring action driven by internal and external factors. When we first encounter an issue, we study the set of inputs-outputs at its root, using a model encompassing as many processes as possible. However, hidden ramifications and feedback loops govern biological systems, not just linear cause-effect relationships. Although many variables change over time, everything repeats in recurring patterns.

Applying this "systems-thinking" approach to orthodontics has yielded significant insights, as it has in other fields like finance and urban planning. It challenges the traditional view that addressing facial issues requires focusing on individual measurements or calculations. Instead, we have learned that the overall behaviour of the teeth and jaw system (dental-skeletal behaviour) is a complex interplay driven by interconnected components. For example, think of the face as a network where each element influences and is influenced by its neighbours. We must focus on something other than individual teeth or bones that seem problematic to make a difference. Analysing the network through computational models can identify underlying patterns and connections that govern the system. This "map of interconnections" reveals the hidden causes of visible problems, allowing us to address them at their source, not just on the surface.

This deeper understanding also allows us to anticipate potential imbalances before they develop and adapt treatment strategies accordingly. Just like predicting traffic flow requires understanding the intricate network of roads and intersections, understanding the dynamic relationships within the facial system empowers us to guide its growth towards harmony and function. Because "an organism is in itself both cause and effect" [2].

When a treatment does not work as expected, instead of simply asking, "What went wrong?" we should ask, "What was the situation before treatment?" "how long did the problem develop?" and "What are the underlying patterns and relationships influencing the affected structures?". Our current approach often involves breaking down facial issues into individual measurements and angles, treating them in isolation rather than considering their context and interplay within the entire system. However, this only works well for complex biological systems like the face. Think of a jigsaw puzzle. Each piece has a specific place, creating a static, unchanging solution. Solving it requires patience and logic, and the number of pieces limits the complexity.

In contrast, biological systems like the face exhibit "dynamic complexity," where elements constantly interact and influence each other. Imagine a puzzle where pieces can move and create different combinations. Static analysis would not help here. Even with just three elements like x, y, and z, they can influence each other in various ways, creating many potential states. With numerous elements in the face, the possible combinations are countless. Such an effect highlights the limitations of traditional approaches and emphasises the need for a broader understanding of facial structures and their interconnectedness. Instead of focusing solely on individual aspects, we must consider the dynamic relationships and historical context to address complex orthodontic challenges effectively.

Studying a complex system like the human face requires more than simply breaking it down into individual components. This "reductionist" approach misses the bigger picture, overlooking the dynamic interactions and historical context that shape the system's behaviour over time. No element exists in isolation; neglecting this interconnectedness leads to incomplete understanding. Quick fixes targeting isolated issues within complex systems have proven short-lived and ineffective in various fields. Further complicating matters, these systems exhibit "emergent properties"—qualities that arise from the organised whole, not present in individual components. Solely studying individual parts cannot explain the rules governing these properties; they emerge from the system's overall structure and the interactions between its elements. In essence, a part within a whole behaves differently than in isolation.

In orthodontics, more than traditional methods based on clinical and radiological data are needed to capture the whole picture. New theories are required to handle the intricate "organisational information" embedded within the system. Moreover, the interaction between the facial system and external factors like breathing patterns further depends on internal relationships between its components. For example, the effects of oral respiration will differ depending on whether a patient has a specific jaw structure (maxillo-mandibular hyperdivergence or hypodivergence). This dependency underscores the complexity of living systems, where the same action can have diverse consequences based on individual variations and timeframes. While treatment might address a specific issue, it might not solve the underlying problem. Destabilising forces within the system can lead to relapse when ignoring the bigger picture. The intricate details of individual components alone cannot capture the dynamic complexity of facial systems. The challenges and paradoxes in treatments highlight the need to move beyond reductionist approaches and embrace a more holistic understanding of the interconnected relationships and historical context that shape facial development and health.

For orthodontists, a significant challenge lies in the delayed and indirect nature of treatment outcomes. The consequences of actions taken today might manifest much later and in areas far from the initial force application, making real-time evaluation difficult. Furthermore, the traditional "transfer of knowledge" approach, where human experience guides machine learning, might be limited when dealing with inherently complex systems where even small changes can cause widespread ripples. Our tendency to think in linear cause-and-effect chains—A is leading to B and B is leading to C—does not capture the complete picture of complex systems like the face. Every action is cause and effect, creating a loop of interconnected influences. Such a loop raises critical questions: Where does the problem truly originate? Where should we intervene to alter the system's overall state? Moreover, are we sure our starting point is valid?

Since a system persists through the interactions of its components, it is constantly changing, making static analysis inadequate. Focusing solely on linear, analytical solutions often leads to symptom-based treatments that do not address the root cause. In these dynamically complex structures, the connections between elements are paramount. Paradoxically, sometimes facilitating a root's movement within the jawbone requires reducing the force applied to the crown. Additionally, individual variations and changes occur gradually, not instantaneously. Is it the external context that drives change, or do inherent features within the system dictate its evolution? Analysing digital orthodontic data, we can create an "evolutionary context" that sheds light on individual characteristics. Changes in variables with low variability (meaning they are consistent within an individual) play a crucial role in determining how the system develops among various possibilities. These compatible features, known as "traits," are fixed over time (e.g., gender) and act as reliable starting points, while more variable characteristics ("states") represent fluctuating aspects that can change over time [3].

In essence, systems thinking encourages viewing the face as a unified whole, shaped by the intricate relationships between its parts. Orthodontic treatment modifies the previous natural geometry of the tooth fit. When we apply a device that does not work we are at the mercy of the patient and his family. We need to figure out quickly what was wrong with the logical reasoning and why our logic did not coincide with the patient's morphogenesis. Systems thinking analyzes the structure of the dentofacial complex, trying to define the fundamental structure of the system, and what role each element plays in holding the structure together. When a computational model accurately reflects a patient's unique characteristics and self-organising processes, it can offer valuable insights for other patients and situations. Orthodontists can unlock clues about future development based on present features and past experiences by combining systems thinking with computational tools. This combination allows for a shift from treating isolated issues to addressing underlying systemic causes of disharmony. Anticipating the downstream effects of interventions empowers clinicians to prevent problems before they arise rather than simply reacting to them. While perfect prediction in medicine remains elusive, even small advancements in understanding complex systems can yield significant benefits. Each step forward in knowledge paves the way for more effective and personalised treatment approaches.

In complex systems like the human face, pinpointing the cause of a problem can be challenging because the root issue might disappear when the visible effects manifest. This feature also applies to the time dimension, with problems sometimes taking a long time to become evident. This delayed nature makes it difficult for orthodontists to connect current observations with past events, making perceiving and understanding these delays a crucial skill. Instead of rushing to immediate solutions, orthodontists should consider that the lingering effects of previous interventions still influence the system and that the proposed interventions might interact with or amplify past treatments. When assessing a new patient, the orthodontist arrives after the issue has already developed, lacking insight into the gradual changes the facial system has undergone throughout the individual's life. This "latecomer's disadvantage" means they might miss the subtle but persistent effects of past problems, growth spurts, or treatments on seemingly unrelated parts of the face. Because space and time can separate cause and effect, our ability to learn and adapt falters when dealing with issues arising from distant points within the system. The root cause of an undesirable treatment outcome might lie far from where it appears. As a result, the "experience" we transfer to machines for analysis is often incomplete or even misleading, as it relies on partial data and potentially flawed interpretations. This feature highlights the need for a more comprehensive understanding and consideration of potential delayed and interconnected effects before implementing orthodontic interventions.

In complex systems, the problems often worsen over time. Overlapping issues can accumulate, sometimes focused in specific areas, but cause varying reactions across different regions. Tracing the visible problems (effects) back to their underlying causes (dynamic processes) is challenging. However, these complexities are only partially beyond comprehension. Within these systems, often underestimated connections run deep, subtly influencing outcomes. Due to these unseen processes, interpreting single-variable analyses (one cause, one effect) with caution is crucial.

Similar to how seemingly harmless factors (but interacting!) can trigger an epidemic, so interacting facial features co-vary over time (multivariate analyses) and determine dentoskeletal growth. Donald Enlow's work [4] linked individual cephalometric measurements to their skeletal counterparts; similarly, a systems approach views each feature's value as emerging not from a single source but from the interconnectedness of the entire network. Localised areas that trigger problems, like growth spurts at the wrong time, are also influenced by the properties of all other elements. As a living system, "…everything is the consequence of everything else…" meaning each part in some way "contains" all the others. This understanding emphasises the need to move beyond isolated analysis and consider the intricate network of interconnections within the facial system. By looking at the bigger picture, we can gain deeper insights into the complexities that challenge traditional approaches and develop more effective solutions for facial issues [5]. When they speak of 'shared information' as the basis of mathematical and formal reasoning of complex systems, computer scientists formulate concepts that are not significantly different from those expressed by the wisdom of the ancient East. The algorithms discover regularities in the data flow, noting that certain parts have common characteristics in the dentofacial ecology. Computer scientists compress these regularities to predict the future. The regularities derive partly from general laws and partly from accidental events.

The behaviour of complex systems might stem from simple rules [6]. Such a property drives orthodontic researchers in their search for the general rules underlying the evolutionary ramifications of the facial system. The hope is to reduce uncertainty about their progression over time and be able to speculate more thoroughly on their global 'emergent properties' not shared by any of their constituents [7]. In orthodontics, the temptation is to see similar general laws from the superposition of often enormously different individual evolutionary possibilities. As stated, not all patients respond similarly to orthodontic treatments: the success of early intervention depends, at least in part, on previous and subsequent craniofacial growth and adaptation [8]. Many factors may contribute to the treatment's success, but not all of them are visible: among them, the type of growth and the ability of the treatment to effectively influence growth. It takes work to identify which process should be attributed to which system element. Patients with severe skeletal malformations will not receive much benefit from the orthodontic treatment. Suitable treatment must stimulate self-correction; thus, patients who have already exhausted their dentoalveolar compensation abilities before the growth spurt will receive little benefit.

2.2 Understanding Facial Development and Malocclusion

Beyond individual growth patterns and specific defects, orthodontists must consider how the facial system adapts and maintains its overall structure over time. This requires flexible approaches based on real-time data rather than predetermined solutions. Visualising the "connecting structure" is crucial to understanding a patient's past development. Similar to how a seashell's spiral reflects the mollusc's growth process, facial structures reveal clues about past adaptations through principles like symmetry, repetition, and modularity. By studying these patterns, we can gain insights into how the individual system has addressed challenges in the past. For example, analysing patients' facial structures with a specific jaw misalignment (Class III malocclusion) revealed three distinct groups based on their growth patterns [9]. In each cluster the craniofacial morphology imposed specific growth constraints, and specific endogenous reinforcing circuits. These clusters create their behaviour over time according to their skeletal structure. The cluster assumption is that underlying the diversity of possible morphological events, a limited number of recurring patterns and generic structures in each cluster help circumscribe a correct interpretative insight about the system progression—more on this critical topic in the following chapters.

The systemic thinker, assisted by computational algorithms, knows how to find connections with elements functionally connected to the variable under investigation and what kind of influence they exert. The paradoxical simplicity of the laws of complex systems also displays probabilistic elements underlying their dynamics. We will see examples of why studying the global relational order of the facial system instead of the specific details is advisable. The macroscopic orthodontic world is better defined. In statistical physics, studying the global behaviour of a gas (the 'laws of state') while neglecting the behaviour of billions of individual molecules is convenient. Reasoning about the average behaviour of gas molecules means significantly reducing the 'phase space' of the system and the related predictive complexity [6]. Even in biological systems, we can observe a supramolecular coherence of millions of molecules, tissues, and organs that realise only one potential of the many possible solutions.

Looking at a system as a whole means automatically highlighting several mechanical constraints and dynamic needs. We define it better because it has become less diversified, and logical contradictions emerge. Reasoning by systems orthodontic concepts, we know that in 80% of the growing subjects, the overall facial pattern, formed by the proportions and profile, remains stable over time [10]. Applying itself to the facial growth predictions, the orthodontic operator experiences the gap between the logical-clinical premises of the model and reality. Statisticians call the errors 'residuals' of linear regression. They stem not so much from the detection mechanisms' inaccuracy but from the orthodontic matter's inherent randomness and the enigmatic variations during growth. Such randomness is precisely what the orthodontist tries to compress.

Computational procedures learn and adapt to the data structure [11]. Probabilistically governed biological events tend to assume an orderly, simple behaviour in the long run, an approximate regularity, whose system macro-variables constitute the statistical information missing in the micro-variables. Algorithms can allocate a probability distribution based on the behaviour of each variable and aggregate the probability distributions on the system's global behaviour. Thus, it is possible to understand complexity through probability, starting from the macro simplicity to decipher the micro complexity.

A qualitative, non-formalised language is sufficient to understand that a forest responds to climate change predictably, whereas an individual tree may have an unforeseen adaptation. A system's local and global dimensions are not separable. Local phenomena of morphological distortions can produce global damage and vice versa. Defining the whole structure of the network is a complex problem of nosology. In Chap. 3, we will see the advantages of morphological coarse-grain pre-classification of patients ("fuzzy" cluster analysis), integrated with topology from Network metrics, in consistently decreasing uncertainty about facial growth progression and treatment effects.

Like all complex systems, the craniofacial consists of various smaller systems (subsystems), each with its way of functioning. When it comes to misaligned bite problems (malocclusion), networks within the face can have smaller subnetworks ("modules") where closely related features are highly interconnected [12]. This property suggests that misaligned bites rarely stem from a single cause but from numerous interacting biological processes affecting pre-existing facial features. Malocclusion is rarely a simple consequence of a single gene product but rather is the trace left in the facial morphology of multiple pathobiological processes that have interacted with a pre-existing network of facial features. During the growth process, tension phenomena appear in some skeletal components as compensation mechanisms for neighbouring deficiencies: the phenomenon of disposal and uniform distribution of feature defects within biological systems, the "Constructal Law" explained below.

In maxillomandibular hyperdivergence, masticatory tension tends to thicken in the molar and premolar areas, whereas in hypodivergence, the tension spreads towards the incisor area. Since tension is not an integral part of the structure, to channel it, the system tends to organise itself into biomechanical tension patterns, in which the centre (the support) tends to move towards emergent centres of balancing of tensions (displayed in the already mentioned modules). Suppose dentoalveolar renormalisation mechanisms fail to mitigate the effects of local muscle and joint tension. In that case, these areas can become emergent morphogenetic patterns that tend to replicate and spread themselves as growth attractors. These areas display the highest probability of future growth events and morphological structure.

2.3 Craniofacials Complexity

Understanding the past can inform how we influence the present in complex systems like the human face. Facial disharmonies, for example, hold valuable information about past interactions within the system and how these patterns have propagated. Techniques like network analysis help visualise these interconnected pathways, allowing treatment to target key points ("hubs") that can influence multiple areas simultaneously. Similarly, Bayesian networks offer a simplified representation of interacting components and their influences, providing clues about where to initiate treatment for optimal impact. In the network analysis, we model the various areas as nodes (hubs) that send information to neighbouring nodes according to a topology of preferred interactive circuits. An unbalanced system must exert more global tension among constituents to keep the parts under control. The muscles surrounding the area of imbalance contract to support the out-of-frame part in place. Chewing tends to localise in the most active regions, following the contractions that lead the connective tissue to adapt to the path of least resistance for the masticatory function [13]. Overall, the unbalanced system tends to become more rigid. The dominant parts may inhibit some parts of the system and may lose part of their functions.

The relationships among nodes of the facial network tend to be nonlinear, so the effects tend to be disproportionate to the causes. We shall see in detail in Chap. 6 that network analysis highlights these situations by describing them using a metric made up of closeness, stress, shortest path metrics, and others. Treatments conducted without awareness of these complex inner aspects are at the root of the release of

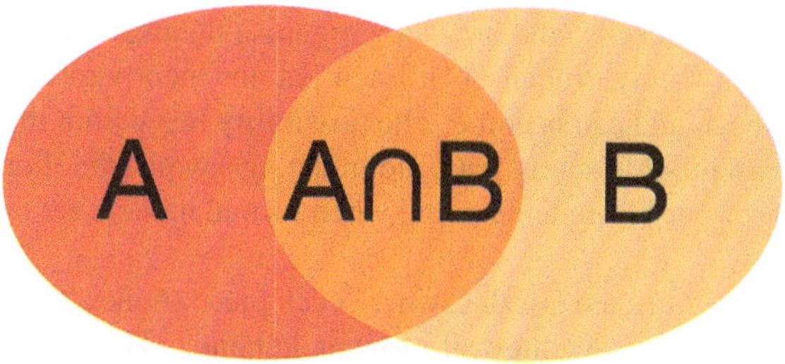

Fig. 2.1 Each additional piece of information allows a sub-classification of characteristics to define the patients' situation in more detail. A and B are aggregate pieces of information derived from different angles. The symbol A ∩ B indicates the overlap of A and B. It refers to a situation with the information from both sources, information that the first angle alone could not provide. The overlapping of two different descriptions generates information of a new logical type with greater "depth"

excessive forces that bring the tooth elements out of basal bone support, resulting in crystal bone loss, instability, gingival recession, root resorption, and relapse. The overlapping of several sources of clinical-statistical description allows for additional information. As an example, the PP-PM angle (i.e. the angle that is between the palatal plane and the mandibular plane) can receive further influences about its ability to define the overall skeletal divergence from the co-occurrent dimension of angles between the body and mandibular ramus (ArGoMe angle), the skull base and palatal plane (PPSN angle), the ratio of anterior to posterior facial height (NMe/SGo), and the vertical distance between the incisal (overbite). The actual value of numerical feature information is linked to the relational aspects, so, as mentioned, we must consider connectivity as a biological characteristic (Fig. 2.1).

The growth outcome in the various facial types is influenced by constituents' original size and ratios (in complexity theory, the "sensitivity to initial conditions") and by the accumulation of reciprocal attractions and repulsions. The orthodontists see these dynamics as local accumulations of closely related characteristics.

Living matter organises itself according to laws and principles of complexity, realising the potential contained in the laws of physics to create the morphologies of life. The rules of thumb of orthodontic living matter say that the head-to-head molar relationship at six years of age becomes a first-class relationship only in 75% of cases. Moreover, significant changes can also occur in the last growth period (after age 15). Overall, 80% of growing subjects keep their facial pattern the same. These are poorly discriminating rules, from which it is difficult to make predictions about the individual patient's future.

According to John Mew's theory, each cell can get information about other cells' activity and the system's ultimate objective. "......if a tooth comes into contact with some adjacent structure with a different positional information, the cell activity slows down, possibly because of conflicting instructions" [14]. Therefore, the control of growth is present in the cells themselves through their ability to respond to positional information from cells around them. The genetic information is inside the cell, but the contact between the cells, and how they organise themselves within a particular environment, is ultimately crucial [14]. Each patient incorporates what he is now and what he has been and can become. The possibility lies within the characteristics that computers can provide a pragmatic aid in the jigsaw puzzle of options. But they can also be a reason for misunderstanding when formal models are transferred to the patient incorrectly.

Our faces are like intricate puzzles, where each piece (bones, muscles, etc.) interacts with others, shaping how they all grow and change. This "new order" created by these interactions makes form and process inseparable. While orthodontists can nudge things by targeting specific areas, we still need a complete understanding of how the face organises itself over time. It's like trying to solve a puzzle without knowing the bigger picture. Instead of focusing on individual features of the face, orthodontists should target the connections between them. Think of it like a network, where changing one node can have ripple effects. By identifying critical connections and "imbalance pathways," treatment can aim to harmonise the entire system. This approach leverages the natural interconnectedness of facial components, shaped by past growth and potentially unpredictable interactions. It's like exploiting the delicate balance within the face, considering its history and the "emergent" changes during development.

We will see that the central metaphors of "complexity" and "thinking in systems" are the network, with craniofacial features represented as nodes and feature interactions defined as links. Just as in the plant ecosystem, in which adaptive constraints force plants to establish strategies to efficiently use limited water resources, light and seed dispersal, competitive strategies between dental elements occur during the permutation to grab space in the arch. Transferred to orthodontics, the results of these strategies define relatively simple models that examine phenomena at different levels of organisation and with other grains: overall harmonious/imbalanced growth, sufficient/insufficient arch space, vertical/horizontal imbalance, muscular hyper/hypotony, and so on. Observing coarse-grained biological systems is a stratagem for simplifying and handling the vast complexity of hundreds of components, revealing global emergent properties and unlikely skeletal progressions [15]. Following the theory of spatial and temporal dynamics (spontaneously arising in plant communities), orthodontists consider malocclusion as a form of adaptation to dental elements' simultaneous use of space and time. As in plant ecology, the succession and zonation of dental constituents result from the same basic processes: the effect of a temporal shift in dominance processes combines with a spatial change in dominance processes between the various strategies and functional roles.

By the end of the seventeenth century, the study of mechanics as an experimental science, following in the footsteps of the discoveries of Newton and later Euler, Lagrange and Laplace, had led to a representation of the world as a perfect mechanism. Supported by the mathematical tools of differential and integral calculus and the experimental conclusions provided by celestial mechanics, the evidence led scholars to conclude that every event happens because of a precise cause or a chain of events with sufficiently determinable details. The philosophical consequence was that knowing the causes that led to an event allows us to infer the details of the subsequent events. During the twentieth century, Newtonian mechanics was extended and improved through two major ground-breaking theories capable of delving into the dynamics of the infinitely large and the infinitely small: the theory of general relativity and quantum mechanics. But a few years earlier, Darwin's studies of evolution and the advent of the theories of thermodynamics challenged the Newtonian monopoly on the laws of physics. The concept of entropy- the tendency of closed systems to irreversibly increase internal disorder- promoted the idea that Newtonian mechanism could not fully explain the complexity of phenomena related to changes in the forms of energy inherent in physical systems, especially in living systems. Toward the end of the twentieth century, complexity theory proved to be a useful cognitive tool for penetrating complex dynamic phenomena still lacking an explanation, phenomena for this reason generally referred to as "chaos." This Greek word identifies the inability to understand the elusive limit that prevents complete observation, especially prediction, about the evolution of an event. While the Newtonian world works incredibly well in describing simple systems, many complex social, economic, and physical phenomena (such as those involving meteorology) need an entirely different approach.

In biology, the cell provides an example of a complex patchwork of integrated tools formed by thousands of interacting proteins, each of which can perform a specific function. Each cell possesses the ability to monitor continuously itself and the environment it lives in and to adapt. It can field different protein combinations: when it senses the presence of glucose in the environment, it puts the necessary protein combinations in place to accelerate its metabolism. When a cell is damaged, it provides the essential proteins for repair. This ability to decipher information and produce specific proteins derives from a complex network of transcription factors, which makes the cell capable of sensing signals of a physicochemical (osmotic pressure, temperature), biological (signals from other cells, toxic molecules), and nutritional nature. Each transcription factor can rapidly transform from inactive to active in response to a signal and activate a specific battery of genes, which are transcribed into messenger RNAs translated into proteins. The network of transcription factors constitutes a formidable internal representation of the environment, an intracellular mould endowed with more than 300 degrees of freedom that can accurately reproduce the complexity of the external environment within the cell. This internal representation enables the cell to perceive and solve problems, reacting in an activator or inhibitory manner depending on the environment. These "cellular machines" behave as if they have a conscious purpose. That is, they calculate.

We find capabilities on different scales in tissues, apparatuses, organs, organisms, and ecosystems. Cellular differentiation, hystogenesis and morphogenesis of skeletal elements become enveloped in networks of informational macromolecules to convey information between the mesenchyme, epithelia, tissues, and organs. Numerous interactive molecular processes of synthesis, storage, release, and inactivation of extracellular organic matter are at work during dentofacial morphogenesis. Chondrocyte membranes are involved in complex epithelium-mesenchyme interactions daily in many embryonic tissues: skin, hair, salivary, thyroid, pancreatic glands, and others. During the orthodontic treatment, the periodontal tissue responds to the forces with a cell explosion. Adaptive processes depend on an interaction between general phenomena operating as constraints ("laws") and the locality, variety, and spatiotemporal singularity of events ("randomness") [16]. Stochastic processes add inherently random elements in the dynamic evolution of complex systems: in the craniofacial system, the interaction can be interpreted in a physical sense as spatial proximity between skeletal elements, in a biological sense as the attraction of forms, in a morphological sense as forms collisions, in a temporal sense as dental competitions during the exchange, in a computational sense as any information-sharing situation between elements during morphogenesis. Each of these dimensions follows specific laws.

2.4 The Craniofacial Apparatus

Orthodontists model the craniofacial apparatus as a functionally organised structure, but the organisation is outside the visible. Therefore, features appear as signs of an underlying invisible organising structure. There is not just one thing but a spectrum of things in everything. Malocclusion is a non-random distribution of dentofacial defects: the same internal structure organises the system's kinetics during the facial growth process, favouring specific collisions and interconnections of some features and the exclusion of others. We can determine the results in the most severe forms of morphological imbalance, over-channelled behaviour that serves no visible purpose.

Patient-related cephalometric variables correlate with each other to varying degrees. As it is almost impossible to ignore the statistical effects of global interdependence, orthodontic operators should consider the data's intrinsic multidimensionality to avoid procedural weaknesses related to the phenomena we will discuss. Skeletal angles with values within the limits of the average but strongly linked to surrounding skeletal unbalanced elements can be considered "reactive". Likewise, the additional importance of information from a lean site close to a growth centre, thus influencing and translating neighbouring elements, should be considered appropriately.

Techniques for framing the behaviour of the cytoskeletal system over time cannot formulate detailed quantitative predictions on future evolution. Still, they can describe some statistical macro-properties and hope to predict something qualitative. In our data concerning more than one hundred Class III patients following longitudinally without therapy, we have observed that all subjects with early severe sagittal imbalance (CoGn/CoA > 1.47) continued to worsen over time, while patients with more nuanced Class III characteristics, in 30% of cases tended to stabilise the imbalance, or even [9]. We will see how local skeletal information behaves about its surroundings and how the same global imbalance acts as an organising structure with self-replicating effects and endogenous binding specifications. Emergent disharmonic properties result from local member-to-member interactions outside of a master plan of action or overseeing authority; the impact extends to the whole system, following preferential paths, partially decipherable.

As the mathematician Richard W. Hamming stated, the purpose of computing "…is in insights, non-numbers". Current digital instruments allow us to follow the temporal dynamics of many phenomena and capture relationships and behaviours between causes of variability in temporal biomedical data. Complex biological systems can move from a state of order to disorder, and vice versa, and change their growth trajectory without any apparent external intervention, thanks to a capacity for global inside information processing that is not present in any component. At the macroscopic level, emergent reactive qualities are foreign to the microscopic elements: propensity, unfolding, anticipation, self-organisation, prediction of environmental change, and taking of countermeasures before the actual change occurs. This emergence comes from biological computation intrinsic to living matter, with specific characteristics we are beginning to learn about. The various interactions of the parts often result in simple, decipherable laws and patterns of behaviour, whose evolution goes through a sequence of micro-states. Each probability depends on the macro-state of the system, in which parts interact more closely with their closest neighbours. The laws of a system's macrostates derive from simple probabilities concerning microstates (Strevens, 2004). We stated that the overall balanced craniofacial pattern that the individual presents at a young age will be maintained, with 80% probability, in adulthood through continuous widespread processes of self-organisation and balance. Let's see some details.

To understand self-organisation, one must understand the importance of schema. Development occurs within many interrelated morphological aspects that make up, as a rule, a balanced overall structure. Harmony is based on parameters that approximate the population average, matched by age and sex, and allows for the expression of average facial expansion and change, according to the principles enunciated by Horowitz and Hixon in the 1960s [10]. The facial pattern represents the "frame of reference," the room for manoeuvre on which to build the corrective orthodontic plan. From 10 to 12 years of age onward, there is relatively little facial growth compared with previous growth: in subjects undergoing orthodontic treatment, skeletal changes before and after this age are mainly due to increase, not treatment. For subjects with harmonious facial proportions, sufficiently accurate growth predictions can be made at 12, comparing the values found with the average values for the population.

In a complex system, the concept of the average means a state of harmony achieved through the multiple connections of the disharmonies of the parts. In the presence of a high variance of components, e.g. when there are many patients with extreme morphological characteristics in the dataset or when the time interval in the various surveys is too wide, the average may not sufficiently express the central tendency of the system. When there are too many dimensions and individual variations in the timing, duration, and intensity of growth, morphological similarities no longer apply [17]. The same cluster's patients subsampled differently, may express different averages and different rules despite the apparent similarity of facial characteristics. We do not know the extent and magnitude of what we ignore; there may be unknown, clinically undetectable variables that heavily influence output. But we don't see them.

Computational predictions help to understand whether specific predictors are correlated with each other so that other predictor variables can explain their apparent effect. A reliable prediction combats the risk of taking incorrect or untimely therapeutic actions based on insufficient, inaccurate, or erroneous knowledge.

The predictive analysis process generates a model by examining the relationship between the predictor data and the outcome. We make the best use of what we know about the patient in the form of data to obtain more and more precise probabilities about what will happen in the future. However, the relational complexity between the components of a biological system can make the definition of general, reliable, and reproducible laws problematic.

Every orthodontist knows that general rules accompany the evolution of the profile in middle and late adolescence. For instance, toward a general flattening due to dominant growth of the chin and nose, along with the decrease in the relative prominence of the dentition within the profile. Orthodontic rules have exceptions, the reasons for which we do not fully know. Position anomalies, dental ectopias and diastemas can manifest in mixed dentition and sometimes correct spontaneously later. They may underlie unexpected problems involving nearby structures, altering the dynamic nature of dental adjustments. It is challenging to separate general statistical phenomena from biological phenomena related to growth, physical phenomena generated by the mechanical collisions of components, and the emerging random phenomena of complex systems.

2.5 Craniofacial Complex Evolution

Biological laws do not necessarily follow simple thermodynamic laws because they are open systems out of equilibrium. Network theory allows the modelling of the emergence of unexpected progression over time of many living structures, delving into the combinatorial properties and internalised competitions of components, not in purely mathematical terms but also in terms of specific behaviour emerging when dealing with concrete developmental issues. For this reason, it is convenient to plan the interceptive intervention by freeing the system from the obstacles to the dynamics of the occlusal adjustments, with priority on the shape anomalies of the arches and the global skeletal defects of the facial system; secondly, addressing the propagation of defects by following the flow from sources to the sinks, facilitating disposal systems of defects (Fig. 2.2).

If we superimpose a grid on image b (Fig. 2.3c) we notice that the figure is 'cut' into many smaller images, each contained in quadrants of the same size. It is a way of imposing a spatial longitude and latitude, to highlight the 'disorder' of the facial image. We can more easily visualise what is outside the normality square to the right, to the left, up and down. It is a way of harnessing the irregular processes of nature. We can classify disorder into particular constructs, but this is only one way of conceiving the world. The straight lines and angles of cephalometry impose a logical scheme on an organic world in disarray, impose an order on chaos. This sounds like a lot, but it is probably only half the story. In reality the image is never divided, there are no separable parts. Each component of the facial bones has boundaries in common with its surroundings, no component has intrinsic properties that belong to it alone. It indivisibly shares all properties with the components with which it interacts. Therefore, it is difficult to identify the best focal point for analysing the system. Because dental arcades continue to change even after age 12, tooth-arch balance is another critical element in conditioning therapeutic intervention. Once full cusp-fossae gear appears in the developing dentition, it is unlikely to change later. This phenomenon happens in subjects with normal or acceptable craniofacial features: from patients with severe disharmonious facial patterns, we should expect different constraints, different teeth collisions, growth trajectories, and different structural growth forces. And even more so from subjects with truly malformed craniofacial features.

Figure 2.3 shows many interacting factors that make learning hard for a machine. To simplify, we can group variables and focus on data connections rather than making assumptions specific to only a few patients. Finding similarities between patients helps uncover shared patterns, which are easier to learn than individual details. However, finding "general laws" in orthodontics is challenging because each patient is unique. Therefore, building models might be complex. Morphology constrains growth: one grows within one's structure. Among teratological patients, we observe a relatively linear non-functional skeletal progression. We will see that the internal generative properties of heavily unbalanced or frankly malformed subjects make them resistant to dentoalveolar renormalisation phenomena. These patients express strong dominance growth relationships and structural rigidities related to compensation phenomena, making the Gaussian probability distribution and the predictive generalisation to other patients problematic.

Craniofacial harmony results from the correct mutual position between the teeth and surrounding structures. A harmonic structure characterises a functioning chewing mechanism, healthy gingival tissues, satisfactory facial aesthetics, respiration, deglutition, and speech. Due to the weakness of general morphogenetic laws, it is possible (in our Class III dataset, in about 20% of cases) that a patient examined for the first time within the age of eight years exhibits a somewhat different whole morphological facial picture at the end of growth. This feature advises avoiding too easy predictions based on the experience of apparently similar cases [20]. However, when the network of features is strongly connected internally, the behaviour of the whole regulates the functioning of the individual variables, each of which contributes to the regulation of the others. Such a phenomenon occurs in the dense internal organisation of Class III malocclusion (thicker than the I and II Class malocclusions) or when a subject exhibits a high membership to a morphological cluster or a severe facial dysmorphism (details in Chap. 4) [21].

When forces of skeletal disequilibrium take over, when there are too heavy competitive dynamics, growth discontinuities, or too many outlier values, the Gaussian distribution of the dataset is lost, as is the meaning of the arithmetic mean. If the bell-shaped curve is skewed, the linear regressive statistical approach is inadequate in defining the occurrence of an event upon the occurrence of another. Throughout

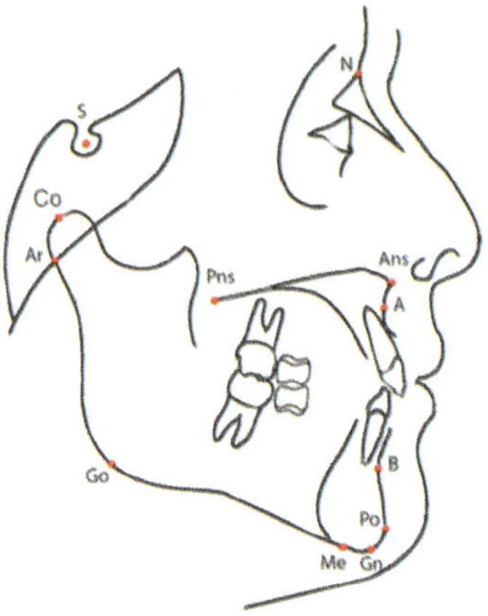

- ***S-N*** *(mm) Antero-posterior length of the cranial base*
- ***S-Ar*** *(mm) Distance from point sella to point articulare.*
- ***NSAr*** *(degrees) Saddle angle*
- ***SNA*** *(degrees) Antero-posterior position of the maxilla to the anterior cranial plane*
- ***SNB*** *(degrees) Antero-posterior position of the mandible to the anterior cranial plane*
- ***Co-go*** *(mm) Mandibular ramus eight, distance between point condylion and point gonion.*
- ***Go-Gn*** *(mm) Mandibular base lenght from gonion to gnation.*
- ***N-Me*** *(mm) Total anterior face height*
- ***ArGoMe*** *(degrees) Gonial angle*
- *Palatal plane to Sn **(PP-SN)** (degrees) Inclination of the palatal plane in relation to anterior cranial base*
- *Palatal plane to mandibular plane **(PP-MP)** (degrees) Inclination of the palatal plane in relation to the mandible plane*

Fig. 2.2 The definition of the cephalometric variables. List from [12] Cephalometric values frame orthodontic reality in an orderly manner. Yet, visible morphology cannot exhaust the complexity and randomness of developmental craniofacial phenomena, i.e., the irreproducible specificity that life can invent. Geometric information complements with biological information. Biological information displays substantial variability in spontaneous internal system mechanisms and external stimuli. Statistics is the most rigorous formal instrument to account for system uncertainty. Since our understanding of the biological orthodontic reality is incomplete, essential information is present in the cephalometric signals. Moreover, the effects of abnormal values under different skeletal morphological conditions are not easily decipherable. Constraints between angles and bony segments can substantially condition growth characteristics and distort statistics, probabilities, and phenotypic final occurrences. The cephalometric data come to our observation accompanied and combined with subclinical signs that we do not see or only partially see. Every orthodontic patient is not only what he is now but also what he will be, what he could be and, of course, what he has been. We cannot perceive the dynamics of the past interconnections that have determined the morphologies. We will see in the following chapters that the data we mistakenly believe we possess can profoundly influence our diagnoses and the outcome of our treatments

the following chapters, many visualisations of logical-mathematical procedures performed on patients with severe and mild malocclusions will be presented, together with the implications towards the prediction of the growth outcome and the response to treatment.

Computational apparatuses attempt to penetrate the orthodontic domain by "learning" underlying laws emerging from collective and individual patient characteristics. After processing a considerable amount of data in the learning phase, regression

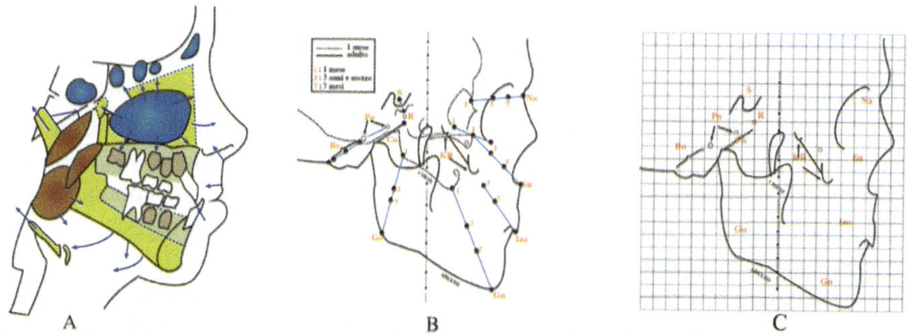

Fig. 2.3 Craniofacial morphology can be considered a complex system characterised by various interactions and feedback. The orthodontist's dream is for linear forward and downward growth but unfortunately, things usually can go wrong. Despite this, we want to frame orthodontic reality in a scheme of underlying order that follows known and reproducible laws. Local regularities cannot permanently stabilise the whole system due to subtle individual differences related to the intrinsic randomness of growth mechanisms, unusual combinations of features, micro-irregularities, and early growth accelerations. Each patient is the unique result of the combined effects of inherently randomly distributed variable processes. Using computational predictive analyses makes it possible, at least in part, to extract signals from the 'noise' of craniofacial development. Combining a series of local micro-details can reveal probabilistic regularities in the evolution of growth processes. Adapted from [18, 19]

models propose as an answer a mathematical formula that binds together the different expressive variables of the situation. Each dentocephalic variable possesses intrinsic geometric multi-dimensional characteristics related to the influence it receives from other specific features. A further situation whereby what is relevant in one patient may not be appropriate in another. The Beta coefficients of the linear regression analysis give the (averaged) effect on the dependent variable Y of a unit change of X, holding all other predictions fixed. However, this may be possible only if all predictors are uncorrelated. Cephalometric analyses fail to report the drifting self-organising properties of the system. However, to assign a weight to each parameter, the regression analysis keeps the value of the other parameters fixed. Statistical procedures adapt poorly to the conceptual drift, which characterises the meaning of each parameter that can change over time, along with the developmental-organisational principles underlying the data. The orthodontist must review the data in subsequent backgrounds and subsequent specificities of the patient's history. This procedure is the orthodontist's delicate preparatory deciphering work before offering the data to the machine (the crucial role of "Feature engineering") [22]. The analysis constitutes the transparent part of the diagnostic logic, while the synthesis represents the implicit operation which has always brought medicine closer to art. Fortunately, the orthodontic field does not contain the additional subjective and random variables of traditional medical semiotics: the symptoms.

Like any doctor, the orthodontist loves the linearity of cause and effect between phenomena. However, some characteristics are so closely interrelated that they cannot be reduced to mere on/off individual variables. The system's nonlinearity stems from the very interdependence of its constituent elements and can cause unexpected and rapid changes. This property creates a patient-specific form of structural uncertainty destabilising because of our limited human capacity to decode information. Thus, simple correlations without causal meaning produce an enormous amount of noise. The mathematics of linearity is simple: the effects of two different causes add up. The trouble is that, in orthodontics, a ten-degree increase in a maxillo-mandibular divergence angle can cause an avalanche of phenomena that modify breathing, swallowing, chewing, phonation, and posture, just as the famous Texas butterfly does.

However, there is no single cause to determine a single consequence; the nonlinearity reflects the global interdependence of the elements of the system that determines the changes. The orthodontist must understand the uncertainty of evolutionary phenomena by entering the error and the fluidity of probabilistic phenomena. The operator does not have to ask the machine, "Tell me what will happen in this patient", but rather "What can I credibly predict?". He would need an answer that expresses the statistical probability that a new patient who has never been examined before can be in a solution area of interest. He would like to know not only whether there will be a satisfactory/poor response to treatment but also whether the chin growth will be downward or forward, whether there will be a need for extraction, and so on [4]. Even if you can narrow the margins of uncertainty, you should not be too ambitious.

Linear mathematics is simple: drawing a line joining two points in space is simple. Using the equation of the best line among data, we immediately know the slope and the point of intersection of the line with the y-axis. We can use this function for several practical matters. In orthodontics, the assumption of linearity implies that the sensors of cephalometry can predict the micro-physics of facial growth and that algorithms can enter patients' subjectivity and guess their morphological future, which could be a reproduction of the past.

Orthodontists know that, in addition to the statistical complexity linked to the nonlinearity of phenomena, there is also an uncertainty related to purely biological reasons. Because there is wide variation in the timing and amount of facial growth before and after adolescence and scarce comprehension of individual maturational times and levels, returning patients to their past behaviour only sometimes seems the most reasonable prognostic strategy. It is possible to understand several actual usefulness from the numerical data of cephalometry by proceeding with systematic order with the help of computerised tools. The orthodontist must follow different pathways: logical (what is the meaning of the data), relational (what is the relationship with neighbouring data), and contextual (what are the constraints to which it was subject during growth). In the end, the numerical data becomes part of a function that contains the most useful starting characteristics, together with the probability of observing in other patients the same way the propagation of defects, the alveolar bone adaptation, and the same spatial arrangements of other facial components.

Induction is the analytical approach that leads from particular facts to general principles. The inductive method, pattern recognition, exploits the model's nonlinear perceptive and cognitive abilities. Uncertainty is unavoidable: the data presented to the algorithm may not be sufficient to clarify it. Even if the data are carefully collected, generalisations about other patients may need more accurate. There are examples from which to extract general orthodontic rules by pure induction. However, argumentative chains of general rules are fragile, as they can take patients who behave differently from expected or contain aspects that are too specific, confusing, or not translatable to other patients. These weak links disturb information transport over time, circumventing the statistical processes of statistical regularisation and standardisation [23].

In an orthodontic clinical trial, individual variability in treatment response is aggregated and compared with all other participants. Skeletal reconfiguration related to the distributed phenomena of bone apposition, resorption and remodelling can be expressed differently in different patients. Data communicate trends, and trends in the facial landscape can change over time: this teaches one to refrain from settling too much into one's ideas (models) when dealing with a reality that admits only fleeting periods of stability and transient corridors of predictability. Statistics presuppose stability and the accumulation of sufficiently comparable examples. Orthodontic patients who are initially similar may diverge in the growth path and may no longer be identical at the end, so the initial similarity can only be apparent. A landmark may change its position (e.g., the Nasion point) concerning other points. Thus, all angles with a vertex in Nasion may have altered values. Nevertheless, the subject has a harmonic profile structure [24]. The virtuality contained in the latent variables was denser in some patients (the 'hidden half' of the facial structure, in Chap. 5), or perhaps some variables had an unequal weight within different patients [25].

The best prediction results arise when the value of a case is an additive function of its attributes. Patients with the most expressive malocclusion features become the dominant cases in the learning set. They are the preferential examples: when an algorithm learns, it continuously finds the best parameters to make the model fit the reality of the input data. The risk of not having visualised the entirety of the players in play advises caution before performing the serial dental extraction methodology, even in situations of discrepancy tooth size-arch size greater than 7 mm. The contours of the problem may change as time passes. It usually gets worse, and extracting dental elements too early can lead to a return to the starting point. It is a decision that should be made based on more than just a one-dimensional assessment of the problem (crowding), considering the whole situation in the lack of space alone.

A virus tends to spread in a susceptible population following exponential dynamics. At a critical point in the atmosphere, one degree more in the temperature may generate hugely different effects. In orthodontics, what is relevant in one context may not be applicable in a different context: a slight difference in a skeletal angle in a critical period of growth can trigger unexpected subsequent changes, activating specific interdependencies between constellations of cytoskeletal features at work. The appeal of mathematics of complexity promises to analyse the effects of nonlinearity related to the co-occurrence of many causal factors underlying natural phenomena. Each of these factors is perceptible, explainable, and controllable, while their interaction can produce enigmatic effects, often spread far apart in time and space. Consequently, seductive explanatory narratives to simplify complex processes can lead to interpretative distortions. It is easy to jump to prognostic conclusions by connecting the diagnostic dots of a patient. When seeing two clinical phenomena occurring over time, orthodontists may identify the latter as the effect of the former (post hoc, *ergo propter hoc*): when noting that the patient tends towards oral breathing we can also note that the patient has deficient transverse palate diameters. The latter is the cause.

It is necessary to frame the patient in the suitable maturation period (in an earlier or later period in which specific sensitive parameters located in critical areas may be the basis of sudden radical changes) to understand their essence. Every orthodontist would like to have such in-depth knowledge of the present phenomena to predict future development scientifically. Predictive models should be tied to a few simple variables, forcibly linked to short time intervals. It is better to reason about co-occurrences, or sequences, of a few critical parameters and, even better, relate each parameter to the overall situation of the system to investigate complex orthodontic phenomena. Orthodontists observe how each parameter relates to neighbours through incremental or decremental relationships and effects.

Each biological evolutionary process contains morphogenetic laws summarising as follows: (1) pattern development; (2) preservation of the pattern when the parameters vary; (3) pattern variation: complexification or impoverishment of the structure; (4) change of pattern orientation. We will see below that, in addition to the phenotypic variability linked to the genotype and the environment, the developmental instability arises from another form of variability: the stochastic component of the total phenotypic variance. Different facial morphogenetic laws can be combined to produce new patterns, as each parameter is affected by the change of many incident parameters. Two skeletal angles of the same size receiving very different pressures from their parameters may result in different end values. Same starting values, different end values, different effects: in the diagnostic process, intuition and acumen are valuable but may only be enough if we visualise all the interconnections and consider all agents.

In the patient's dentofacial structure, the orthodontist detects a morphological outcome resulting from an infinite sequence of competing causes. Relative growth of the maxilla and chin, post- or ante-rotation of the mandible, habits, and growth obstacles related to deep bite are frequent causes of horizontal or vertical imbalance. The enigmatic interactions between phenomena represent turning points during development: the more significant the severity of the dentofacial anomaly, the more it is preserved over time. What lies behind an anomaly can be purely due to the stochastic component (for some authors, from 70 to 97%) [26]. The orthodontist follows the evolution of a characteristic over time and tries to correct it when it seems outside of the acceptable range. They might think that all the other variables remained constant over time, or at least their interaction was constant. Some people believe that the stochasticity inherent to the topic represents only the measure of ignorance regarding morphogenetic processes.

We stated that once an evolutionary orthodontic rule has been identified in a patient, we cannot automatically extend it to other patients because each patient's outcome results from a singularity of interaction and aggregation of disparate measurable and latent individual characteristics. Since each variable can contain a handful of information, the orthodontist would like to obtain a better patient specification from as many sensors and attributes as possible, interweaving seemingly disparate information into something with clinical coherence, predictive capability, and the possibility of correction. The assumption is that better defining the structure can help determine further system properties, such as progression or robustness. Using computational procedures, the orthodontic operator must provide the machine with several "labelled" patients (patients of which it knows the initial and final conditions to offer as examples) to best define the boundaries of the topic and find a function to recognise patients never seen before. Unfortunately, mere craniometric characteristics are often a pale clue of the causes and processes that define the label.

Imagine learning by comparing similar situations. We estimate probabilities by using a few examples to represent possible outcomes. Facial development works similarly, influenced by both individual traits and broader contexts, past and present. We can analyse quantitative measurements using math, but capturing unique qualities can be tricky. Here is the challenge: predicting future outcomes based on current data skips a crucial step. Anatomy is not the sole factor at play. It interacts with other influences differently in each patient, making predictions tricky. Feedback loops create a complex dance between cause and effect, stimulus, and response in living beings. Feeding "sanitised" data to machines might seem efficient, but it can lead to misleading results, hence the "replication crisis" some warn about [27]. The predictive weakness appears more often when it concerns the history (the world) of some patients who have had an anamnestic path full of complex, tangled interactions, which make the numerical reduction of attributes difficult or naïve.

Not only individual but also general orthodontic rules can be subject to contradictory logic: reducing the force exerted on a dental element can speed up dental movement, avoiding the hyalinisation of the periodontal space. Therefore, the uncertainty weighing on the averages of orthodontic values is less than that of the uncertainty weighing on the single value in the individual patient, in the same way, that, as we have already mentioned, the regularities between temperature, volume and pressure equations of state of a gas are defined by simple relations, compared to the difficulty of studying the behaviour of individual molecules.

Thanks to the generosity of many international orthodontic operators, the research group to which the authors belong has collected cephalometric characteristics of a large variety of growing Caucasian subjects suffering from various types of malocclusions. Most of this data (approximately one thousand five hundred surveys) concerns Class III subjects from 5 to 21 years of age. Observations refer to many patients before and during treatment. In Class III, for subjects who (for various reasons) were not subjected to treatment, using different computing resources, we were able to observe some rather unexpected reliefs about the natural history of this malocclusion. Two simple threshold values of two cephalometric angles, PPSN and L1PP (PPSN: angle between the anterior skull base and the palatal plane. L1MP: the Inclination of the lower incisors concerning the mandibular plane), were able to discriminate to a satisfactory degree of accuracy. These subjects, in the absence of therapy, would have seriously increased the Class III skeletal imbalance over time. Subjects with a PPSN angle value lesser than 8.35 degrees and an L1MP angle value greater than 77.1 degrees carried a 22% risk of negative growth. (Auconi et al., 2017). Conversely, in subjects with a PPSN angle greater than 8.35 degrees and an L1PP angle less than 77.1 degrees, the risk of severe growth was 67%. It only takes two parameters with altered values to cause disastrous growth. How come? As we shall see, the contour where we assign these values is decisive. It is quite probable that the definition of the "contour", i.e. the system in which the variables can move, represents a crucial aspect for defining any malocclusion in determining the fundamental importance of variables.

Predictive algorithms can enter the patient's subjectivity by exploiting the values of specific local characteristics when we correctly frame the patient. Complex biological systems connect nonlinear local processes to the global process of which they are a part, reproducing their complexity through the creation of specific local generative rules, which become general regulations about the dynamics and constraints of the process itself.

The systemic logic of local-global interaction helps understand some key concepts underlying the typical behaviour of complex systems. It promotes the development of a new form of nonlinear connective reasoning ("qualitative", "holistic", "decentralised", "relational"). Fritjiof Capra adequately defines "systems as integrated totalities, with unique specific properties that cannot be reduced to simple combinations of the smallest units, nor understood in this way. Instead of focusing on elementary building blocks or fundamental substances, the systems approach places the fundamental principles of organisation at the centre. All these natural systems are totalities whose specific structures derive from the interactions and interdependence of their parts. When a system is broken down—either physically or theoretically—into isolated elements, the systemic properties end up destroyed..." [28].

It is becoming clear that the morphogenesis of many living forms starts from elements of extreme simplicity that follow simple rules to build rich and wonderful morphologies [29]. Accordingly, statistical descriptive and predictive models must be comparable with the multidimensionality of the natural systems. Models must be isomorphic to the delicate relationship between random and deterministic laws, i.e., the evolution of the natural process in which things happen. In this process of natural computation, the global process is the contingent result of more basic local processes that tend to interact nonlinearly and to self-organise in time and space. This specific property must be able to reshape in computational terms the much-quoted sentence: "......in complex biological systems, the whole is more than the sum of the parts".

Stable complex systems need structures that support their processes. Computers show how elements within these systems constantly interact and influence each other. Understanding these systems involves combining detailed descriptions with broader patterns. This property reveals how individual parts, despite their differences, work together as a whole. Within this dynamic interplay, the potential exists for change, like seeds holding the possibility of future forms. During development, shapes combine to create new structures with various properties.

Similarly, it is possible to consider patients as combinations of different aspects, like movements and habits. By analysing patterns and categories, machines can help us understand this dynamic "space between things" where change happens, and forms emerge. Many researchers have sought to clarify the relationships that bind the constituent parts of a system with a whole. This general rule forces its rigour on things by imposing its necessity. The orthodontist derives his prognostic model from a geometry operation of the features involved. The model form must include the prediction, as precise as possible, of the laws governing the evolutionary trajectories of the facial shape to be effective. By defining the nuances of individual contingency and eliminating the background of indeterminacy as far as possible, the orthodontist favours the matching between the general rule and the singularity of the subject. Yet, as Francois Jullien suggests, particularity resists the generality of the law because something always remains "... more or less reluctant to the determination of the "form", so the world is never completely available to the order we want to give it: there will inevitably a gap between the model we design to act and the one we, with our eyes fixed on it, come to realise. In short, practice will betray theory by how little..." [30].

We stated that biomedical micro-processes could be neither suspected nor viewable by doctors, as they are the result of a myriad of interactions to which they did not have access, coming from seemingly unconnected contests [31, 32]. The possibility of having a dataset covering many patients allows us to focus on a system's possible trends and tolerate a specific inaccuracy in some data. Although it may seem extravagant, contexts, an inevitable imprecision in data can improve the analytical capacity of computer procedures and enhance the possibility of generalising predictions over a more significant number of patients. We will see the reasons for preferring information that is not too precise but conducted on many patients over exact information but fewer in number.

Analysing two-dimensional cephalometric images holds promise for understanding patients' facial structures. However, more than simply comparing individuals to population averages is needed. Capturing their unique characteristics is crucial. Case-based reasoning algorithms address this need. These algorithms analyse patients' cephalometric history, essentially "travelling back in time" through their facial morphology. This historical context enriches future growth predictions by incorporating past influences. Unlike traditional deductive or inductive reasoning, Deep Neural Networks rely on stacked layers of data analysis. This process discovers hidden patterns within existing data, offering additional insights into facial development. Although these patterns provide interesting alternative explanations, they do not necessarily adhere to established scientific rules. These algorithms learn from past patient data to predict future facial changes. They prioritise individual differences in overpopulation averages, revealing hidden patterns through statistical methods. Although their approach diverges from conventional scientific analysis, they contribute valuable knowledge to understanding facial development [33].

2.6 Craniofacial Networks

The graph theory allows visualizing the collective network created by the interactions of craniofacial characteristics of class 3 malocclusion subjects who are at risk of exacerbating craniofacial imbalance with growth. The graph enables mapping the spatial and temporal aspects of the phenomenon. The intangible structure of relationships influences treatment approach and highlights what truly matters. Graphs have shown that, in terms of correlation coefficients, the links holding together disharmonic facial morphologies are more robust than those characterising harmonic structures (see Fig. 2.4). The force we exert on the system is antagonised not only by the mechanical reaction of matter but also by the biological emergent forces of attractive interaction of components, adaptation, competition, sensitivity to initial conditions, and optimisation of forms during the growth process. Computational machines make everything calculable and predictable, allowing orthodontists to convert more aspects of clinical reality into data. In many situations, it is convenient to analyse these systems proceeding by coarse grain, i.e., leaving out the finer details to concentrate on reasoning simply about the dominance of horizontal skeletal disequilibrium over vertical disequilibrium.

Doctors are used to looking for the causes of diseases in the patient's history, signs, and symptoms, in the intersection of habits, genetic predispositions, non-predictable accidental factors, and more. Although causation is often problematically nuanced, doctors' reason by cause and effect, not by correlation coefficients. Algorithms enlighten biological processes and transform facts into something probabilistic, and this response requires some adaptation by the doctor. In orthodontics, no precise standards define the nosological picture of interrelations. Recording information in the form of 'categories', i.e., groups of similar patients from which to extract useful information, offers the opportunity to make timely connections between data and clinical events. Unlike many physical objects in a conventional experiment, objects in a computerised experiment do not get their connection through the laws of nature. They follow the numerical laws embedded in the computer program. Experiments are performed in a hypothetical universe of functions and mathematical constants [34].

Since a complex system exhibits systemic behaviour that emerges from the interaction of the components, and because "the whole is different from the mere sum of the components", it is impossible to understand the system by looking at one or few individual details of the system. The laws of complex systems are still deterministic, but minor imperfections in initial knowledge make it impossible to predict the details of the system's future behaviour. Therefore, orthodontic treatment can have nonlinear, counterintuitive effects or effects on parts far from the point of attachment or delayed in time. The illusion of control can occur when the system's behaviour is seen as the result of our action when, in fact, it is the intrinsic behaviour of the system. To control a complex system, we should have a variety of possible tools similar or superior to the complexity of the system (the "law of necessary variety") [35].

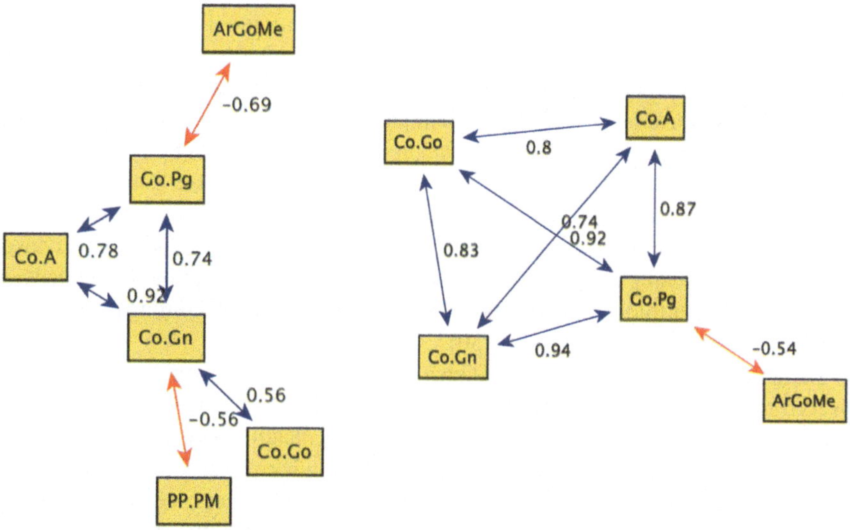

Fig. 2.4 Correlation map amongst cephalometric quantities in 90 untreated class III patients (6–17 years of age). On the left, we have 45 who improved their Wits appraisal during growth. On the right, other 45, who worsened. Note how, on the right, the graph is more compact and has more substantial edges. On the left, the most potent edge is between Co—A and Co-Gn (a sign of more balanced maxillo-mandibular growth). Because of the many interactions between the different nodes, a strongly unbalanced and strongly connected morphological system tends to be refractory to change. Multiple reinforcement recurrent loops facilitate the pathogenic cascades of imbalance. In the course of facial morphogenesis, each cell obeys genetic orders, but also the information it receives from neighbouring cells. The self-organising network of scattered agents creates a higher order. The mutual interaction between the parts determines a recognisable but not necessarily predictable 'emergent' macro-behaviour. Changing a single feature of the system impacts many other nodes, not just the single component. During the growth process, some bone components interact directly with other elements, with others indirectly through interposed components and random collisions. We must act on the interaction network between the parts to carry out a true transformation. The orthodontist's supervision is crucial in deciphering the complex components of orthodontic phenomena to avoid forcibly partial, abstract, or idealised scenarios

At the local level, each defect may have a limited impact. However, the chain of events can lead to a progressive build-up of correlations on a large scale and eventually to malocclusion. Every orthodontist knows how to attribute proper importance to 'external' factors, which are the core of the processes of malocclusion organogenesis: muscle tone, lingual posture, breathing, and viscoelastic structures surrounding defects are all factors that are entirely part of the system. Dysmorphosis is difficult to change because it is not an isolated body within a system but is an integral part of the growing facial system and, as such, has articulated roots and deep connections. Repairing a living system does not mean improving an element but rather rebalancing the relationships that govern all the parts. Possible resistance to orthodontic treatment does not stem from the persistence of a flawed habit or incorrect posture but rather from the system as a whole pursuing the maintenance of an optimised imbalance over time. The biomechanics of the intervention will inevitably generate a reaction/resistance on the part of the system: we consider them from the outset as an inevitable 'emergent' variable built into the system's future. Since they cannot establish complete control over the system and there is no treatment without a cost, the orthodontist must exploit the pragmatism of systems reasoning by putting things in such a way that the system does part of the work. Identifying the "situation potential" is crucial because the malocclusion is often present initially; it is just waiting for the "trigger" to manifest itself. The interceptive intervention must foresee the consequences of actualising the potential, identifying the minimal anticipatory tendencies that will lead to unfolding. "From the subtle to the manifest": becoming can be made predictable "because it is implied" [30].

In the endoskeletal development of post-adolescence, the probable gradually takes over from the possible. Network analysis has shown that, to narrow down uncertainty, one must move from quantities to maps. An exact match between the description and the actual phenomena is impossible in orthodontics. What is interesting is the transition between data and the visible skeletal structure, the processes underlying facial morphology and organisational patterns. In Bayesian network analysis, there is the possibility of unravelling the influences between the various craniofacial nodes (see Fig. 2.5). In the case of Bayesian networks, we can also represent cause-and-effect relationships. Since in subjects who worsen, it is menton B that conditions maxilla point A, it is preferable to focus the distalising therapeutic approach on area B, rather than the 'drawer movement' forward produced by the Delaire mask (the latter, moreover, produces both effects simultaneously). An impact can be transmitted from one node to another, then to another, and then back to the first, causing a cycle of cause-and-effect phenomena, which can alter the predictions.

Modern Bayesian techniques in computational statistics are built on fundamental principles of probability theory to provide a better understanding and visualisation of complex data. Bayesian statistics develops the concept of combining the information in clinical or experimental data with prior knowledge from previous experiments to evaluate the likelihood of specific prognostic hypotheses. Bayesian Networks describe a biological system with multiple interacting components using a simplified representation that captures the interplay between them and how they influence (and possibly be causal for) each other. A popular choice for this representation is the Directed Acyclic Graphs (DAGs), where links represent direct probabilistic dependencies and arrows indicate the dependence's direction. As in living systems, form is the sediment of the dynamics of the underlying past formative processes; Bayesian Network analysis helps visualise the properties of the system in terms of reciprocal influences between the local parties and the whole organisational pattern, i.e., the global properties that no part possesses, which govern the development of organic forms. The communication network of rules of conduct is present in each system component.

Within biological systems, self-generating networks develop with rules that establish the behaviour of involvement of the components. The initial orthodontic diagnosis relies on an overview: there is no single tell-tale; it is a constellation of small or

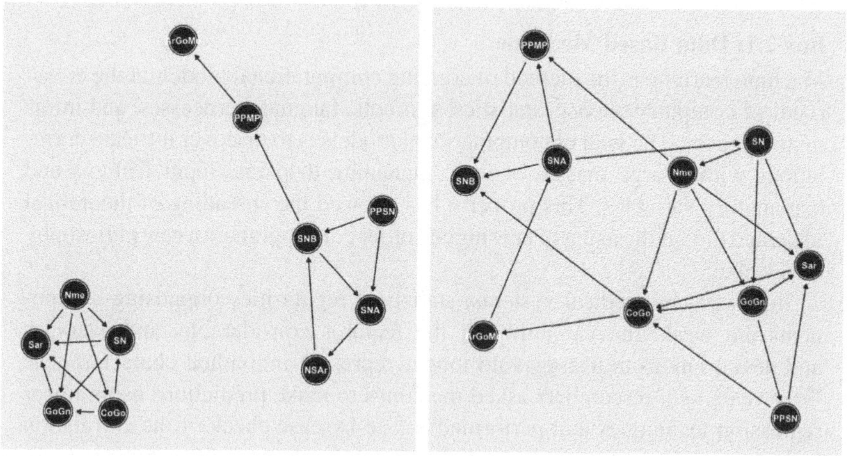

Fig. 2.5 Bayesian Graphs representing the influences between dentoskeletal features in 40 untreated Class III patients incurring growth enhancing the dentoskeletal imbalance (left). On the right, dentoskeletal features of 40 Class III subjects of similar age and gender reduced the maxillo-mandibular imbalance during growth. In subjects whose imbalance worsened, the SNB influenced SNA, while in the subjects whose imbalance was spontaneously improved, we see that SNA influenced SNB. At the bottom left, one can see a thickening of features with dense mutual influences, presumably favouring a shared propensity for development. This thick, self-generating network (module) leads to the coordination of the components and creating a growth attractor. The attractor is a complex geometric model representing the dynamic properties of a biological system; it gives rise to a limited number of possible morphological forms. (Additional information in Scutari M and coll.)

large discrepancies from the expected that sometimes combine to define a malocclusion. Each craniofacial characteristic possesses a multidimensionality: its authority, a specific capacity for representing one's morphological condition, and the relative statistical importance it sends to neighbouring nodes or receives from them. The most influential features amplify their topological effect: they act as network hubs by defining a hierarchy of functional coordination and mutual influences between surrounding features.

Patients with a facial bone structure characterised by early severe maxillo-mandibular imbalance must be placed, studied, and compared in an ecosystem of patients with similar characteristics. The growth of heavily unbalanced patients with a Class III malocclusion is relatively predictable due to several skeletal global and local constraining characteristics: they must be studied following the microphysics of their regularities, as their future will most likely be the reproduction of their past [9]. When adequately characterised, the same altered morphological dimension restricts their morphospace development.

Box 2.1: Data Based Medicine

Machine learning is the method of creating computational models at the crossroads of computer science, statistics, symbolic language processes, and information theory. The goal of computational models is to discover intricate correlations within large masses of data, managing disparate input features and explanatory variables. This property has allowed the spreading of theoretical ideas and the synthesising of new higher-order concepts about a new philosophy of science.

In several biomedical systems, statistical regularities organising components are weak, the variability of the features considerable, and averages and deviations from averages no longer represent individual characteristics. Fifty years ago, researchers asked machines to make predictions using linear regression techniques that performed a case-by-case check of the correlation between two distinct variables, isolated from all other variables in the system. The dream of algorithmic calculations is to rationalise the clinic and relieve doctors of epidemiological research on the prevalence of various disease states. Nowadays, they test all possible correlations between an increasing number of system variables, exploiting the statistical physics of large numbers. Models represent the main application of mathematics and computation in biology and medicine. They result from the attempt to fit data-based and rule-based frameworks with the reality of biological systems. The more crucial the theory explains the problem, the better its generalisation. New information emerges in the mutual accommodation relationships between the system components, allowing us to make predictions. The model must incorporate the primary biomechanical constraints and the logical rules that describe the trajectory of a particular phenomenon, which must be appropriately "engineered" by the human operators.

Computational machines can distinguish what is random from what is regular in the system and consider a patient's morphology as a solution to an underlying dynamic. However, sometimes, mathematical models tend to be unrealistic, as they only feel a few potentially important details. For example, the Lotka-Volterra prey-predator model does not consider the type of prey, predator, hunting territory, the possibility of hiding, and other features. The subtle complexity of many aspects of the patient and the frequency of emergent phenomena capable of changing its evolutionary pathway prompt a computational prediction focused on the system's global behaviour. Moreover, in mathematical reading, statistical reiterations offer answers progressively purged of concrete meanings and tend to formalise disembodied metaphors of clinical phenomena. Physicians must develop strategies to counter calculations based on their discernment, intuition, and experience to avoid the risk of technological elusiveness and paternalism over medical knowledge.

References

1. V.C. Pezoulas et al., A computational pipeline for data augmentation towards the improvement of disease classification and risk stratification models: A case study in two clinical domains. Comput. Biol. Med. **134**, 104520 (2021). https://doi.org/10.1016/j.compbiomed.2021.104520
2. U. Danielsson, Världen själv. Fri Tanke förlag (2020)
3. P. Baltes, H. Reese, J. Nesselroade, in *Life-Span Developmental Psychology: Introduction to Research Methods* (Psychology Press, New York, 2014). Accessed: Feb. 12, 2024 [Online]. Available: https://books.google.com/books?hl=it&lr=&id=k097AgAAQBAJ&oi=fnd&pg=PP1&dq=Baltes+P+B,+Reese+H+W,+Nesselroade+J+R%3B+Life-span+developmental+psychology:+Introduction+to+research+methods.+1988%3BL.+Erbaun+Ass.:+Hillsdale,+N.Jersey)&ots=Ngr20BpmOP&sig=4gGCeo
4. D.H. Enlow, T. Kuroda, A.B. Lewis, Intrinsic craniofacial compensations. Angle Orthod. **41**(4), 271–285 (1971). https://doi.org/10.1043/0003-3219(1971)041<0271:ICC>2.0.CO;2
5. F. Capra, in *The Tao of Physics : An Exploration of the Parallels Between Modern Physics and Eastern Mysticism* (Shambhala Publications, 1991). Accessed: Feb. 12, 2024 [Online]. Available: https://www.google.it/books/edition/The_Tao_of_Physics/NmhfQgAACAAJ?hl=it
6. G. Caldarelli, *Scale-Free Networks: Complex Webs in Nature and Technology*, vol. 9780199211 (2007). https://doi.org/10.1093/acprof:oso/9780199211517.001.0001
7. D.C. Krakauer et al., The challenges and scope of theoretical biology. J. Theor. Biol. **276**(1), 269–276 (2011). https://doi.org/10.1016/j.jtbi.2011.01.051
8. P.W. Ngan, U. Hagg, C. Yiu, S.H. Wei, Treatment response and long-term dentofacial adaptations to maxillary expansion and protraction. Semin Orthod **3**(4), 255–264 (1997). https://doi.org/10.1016/s1073-8746(97)80058-8
9. P. Auconi, M. Scazzocchio, E. Defraia, J.A. McNamara, L. Franchi, Forecasting craniofacial growth in individuals with class III malocclusion by computational modelling. Eur. J. Orthod. **36**(2), 207–216 (2014). https://doi.org/10.1093/ejo/cjt036
10. L. S. Horowitz, E. H. Hixon, in *Nature of Orthodontic Diagnosis* (C.V. Mosby Company, 1966)
11. M. Henrion, Propagating uncertainty in Bayesian networks by probabilistic logic sampling. Mach. Intelligence Pattern Recogn. **5**, 149–163 (1988). https://doi.org/10.1016/B978-0-444-70396-5.50019-4
12. A. Scala, P. Auconi, M. Scazzocchio, G. Caldarelli, J. McNamara, L. Franchi, Complex networks for data-driven medicine: the case of Class III dentoskeletal disharmony. New J. Phys. **16**(11), 115017 (2014). https://doi.org/10.1088/1367-2630/16/11/115017
13. M. Bond, in *Rolfing Movement Integration: A Self-Help Approach to Balancing the Body* (Inner Tradition, 1993). Accessed: Oct. 06, 2023 [Online]. Available: https://www.amazon.com/Rolfing-Movement-Integration-Self-Help-Balancing/dp/0892814446
14. J.R.C. Mew, Factors influencing mandibular growth. Angle Orthodon. **56**(1), 31–48 (1986). Accessed: Feb. 20, 2023 [Online]. Available: https://meridian.allenpress.com/angle-orthodontist/article/56/1/31/56272/Factors-Influencing-Mandibular-Growth
15. T. Smith, M. Huston, A theory of the spatial and temporal dynamics of plant communities. Vegeta. **83**(1–2), 49–69 (1989). https://doi.org/10.1007/BF00031680
16. M. Ceruti, Il vincolo e la possibilità. Accessed: Oct. 06, 2023 [Online]. Available: https://www.raffaellocortina.it/scheda-libro/mauro-ceruti/il-vincolo-e-la-possibilita-9788860302960-1249.html
17. P. Domingos, in *The Master Algorithm: How the Quest for the Ultimate Learning Machine Will Remake Our World* (Basic Book, New York, USA, 2015). Accessed: Feb. 13, 2024 [Online]. Available: https://www.amazon.it/Master-Algorithm-Ultimate-Learning-Machine/dp/0241004543
18. G. Maj, *Manuale di Ortodonzia*, 3rd ed. (1991). Accessed: Dec. 28, 2022 [Online]. Available: https://www.libreriauniversitaria.it/manuale-ortodonzia-maj-giorgio-patron/libro/9788855516846

19. A. Patti, G. Perrier D'Arc, *Les traitements orthodontiques précoces. Quintessence international* (2003). Accessed: Oct. 06, 2023 [Online]. Available: https://www.amazon.it/traitements-orthodontiques-pr%C3%A9coces-Antonio-Patti/dp/2912550246
20. S.E. Bishara, in *Textbook of Orthodontics* (Saunders, 2006). Accessed: Oct. 07, 2023 [Online]. Available: https://www.amazon.it/Textbook-of-Orthodontics/dp/8181470729
21. P. Auconi, G. Caldarelli, A. Scala, G. Ierardo, A. Polimeni, A network approach to orthodontic diagnosis. Orthod. Craniofac. Res. **14**(4), 189–197 (2011). https://doi.org/10.1111/j.1601-6343.2011.01523.x
22. A. Zheng, A. Casari, in *Feature Engineering for Machine Learning: Principles and Techniques for Data Scientists* (O'Reilly Media, Inc., 2018). Accessed: Oct. 07, 2023 [Online]. Available: https://www.oreilly.com/library/view/feature-engineering-for/9781491953235/
23. P. Auconi, E. Ottaviani, E. Barelli, V. Giuntini, J.A. McNamara, L. Franchi, Prognostic approach to class III malocclusion through case-based reasoning. Orthod. Craniofac. Res. **24**(S2), 163–171 (2021). https://doi.org/10.1111/ocr.12466
24. M. Bingmer, V. Özkan, J. Jo, K.-J. Lee, H.-S. Baik, G. Schneider, A new concept for the cephalometric evaluation of craniofacial patterns (multiharmony). Eur. J. Orthod. **32**(6), 645–654 (2010). https://doi.org/10.1093/ejo/cjp152
25. M. Blastland, *The Hidden Half: The Unseen Forces that Influence Everything* (Atlantic Books, 2020)
26. D.L. Lajus, J.H. Graham, A.V Kozhara, Developmental instability and the stochastic component of total phenotypic variance, in *Developmental Instability* (Oxford University Press, New York, NY, 2003), pp. 343–364. https://doi.org/10.1093/oso/9780195143454.003.0019
27. E. Coiera, E. Ammenwerth, A. Georgiou, F. Magrabi, Does health informatics have a replication crisis? J. Am. Med. Inform. Assoc. **25**(8), 963–968 (2018). https://doi.org/10.1093/jamia/ocy028
28. F. Capra, in *Patterns of Connection: Essential Essays from Five Decades* (High Road Books, 2021). Accessed: Oct. 07, 2023 [Online]. Available: https://www.amazon.com/Patterns-Connection-Essential-Essays-Decades/dp/082636280X
29. S.A. Kauffman, in *At Home in the Universe: The Search for Laws of Self-Organization and Complexity* (Oxford University Press, 1995). Accessed: Oct. 07, 2023 [Online]. Available: https://books.google.it/books/about/At_Home_in_the_Universe.html?id=o-Owb5IDkSQC&redir_esc=y
30. F. Jullien, *Traité de l'efficacité* (Grasset, Paris, 1996). Accessed: Dec. 15, 2023 [Online]. Available: https://www.persee.fr/doc/etchi_0755-5857_1997_num_16_1_1257_t1_0155_0000_2
31. G. Caldarelli, M. Catanzaro, in *Networks: A Very Short Introduction* (2012). Accessed: Oct. 07, 2023 [Online]. Available: https://www.amazon.it/Networks-Short-Introduction-Guido-Caldarelli/dp/0199588074
32. A.L. Barabási, N. Gulbahce, J. Loscalzo, Network medicine: a network-based approach to human disease. Nat. Rev. Genet. **12**(1), 56–68 (2011). https://doi.org/10.1038/nrg2918
33. R.C. Deo, B.K. Nallamothu, Learning about machine learning: the promise and pitfalls of big data and the electronic health record. Circ. Cardiovasc. Qual. Outcomes **9**(6), 618–620 (2016). https://doi.org/10.1161/CIRCOUTCOMES.116.003308
34. F. Bertacchini, E. Bilotta, P. Pantano, in *Il caos è semplice e tutti possono capirlo* (Muzzio, 2009). Accessed: Oct. 07, 2023 [Online]. Available: https://www.amazon.it/caos-semplice-tutti-possono-capirlo/dp/8896159164
35. R.W. Ashby, in *An Introduction to Cybernetics* (Chapman & Hall Ltd., 1956). Accessed: Oct. 07, 2023 [Online]. Available: http://pcp.vub.ac.be/books/IntroCyb.pdf

Chapter 3
Seeing the Invisible

An orthodontist's job involves fixing physical problems with teeth and jaws, such as misalignment and growth issues. These problems can be challenging to address because they involve measurable aspects (angles, segments) and less clear concepts like aesthetics and growth patterns. In the past, teeth extraction was standard to achieve straight teeth. Today, orthodontists try to avoid extractions and instead use methods like influencing facial growth, using special tools, and understanding how teeth and bones move. Extractions and surgery are still used sometimes, but only when necessary. Partly, this shift is because extractions can weaken the face over time. Overall, orthodontists today have more tools and understanding to fix teeth and jaws without removing teeth, leading to better patient outcomes.

System thinking provides a new conceptual framework that generates the required measure for the whole organisational mechanisms of biological entities. As Kevin Kelly points out, "…a grain of sand can't produce a landslide, but put together enough grains of sand, and you get a dune that can start a landslide. Certain physical attributes, such as temperature, depend on collective behaviour. A single molecule floating in space doesn't have a temperature. Temperature is a macroscopic quantity referring to a large group of molecules. Science has long appreciated that large numbers behave differently than small ones. The total number of possible interactions between two or more members accumulates exponentially as the number increases. The dynamics of mobs take hold at a high level of connectivity and many members. More is different" [1].

When parents ask dentists to fix their child's teeth, they deal with a complex system. Every part (tooth, jaw, etc.) affects the others. It's like trying to organise a messy room, but where each item somehow interacts with all the others. Dentists can approach this in two ways:

- The mechanical approach focuses on individual parts, like removing weight from a crowned tooth.
- The network approach considers how all parts interrelate, like a "collective reality" with many interconnected functions.

P. Auconi et al., *Systems Orthodontics*, Understanding Complex Systems,
https://doi.org/10.1007/978-3-031-64932-5_3

This network approach connects orthodontics, artificial intelligence, and other fields. It suggests that nature evolves through networks, where different features compete, grow, and solve problems. This property is common in economics, ecology, and even plants. With computers, we can now visualise these connections within an organism, expanding our understanding of its form and function. These connections reveal an essential layer of information, especially in medicine, where "Network Medicine" is gaining momentum [2].

The network structure reflects the dependency structure between the variables, identifying the areas with the highest densities of influential probabilities. Connectivity is one more piece of information to add to the morphological characteristics of each patient; it can be an alternative to formal (mathematical, geometrical) analysis methods. It can define the internal consistency of the facial system and a "space for developmental solutions" dependent on conditions and causalities that cannot be clinically quantified [3]. In orthodontics, as in other fields of biomedicine, there are ungraspable, mobile, and reactive developmental dynamics comparable to those described by Tom Ritchey as "wicked" problems. The connectivity quotient between the components of a system allows for a "General morphological analysis", an expanded form of modelling capable of formulating unique evolutionary scenarios and simulations of side effects of morphogenetic phenomena [3]. Computers use interactive inference models to describe the multi-dimensional nature of a medical problem and parallel scenarios in which patient attributes are as many dimensions as possible. These analyses display the value of a single characteristic. They involve consistency assessment by deconstructing and decomposing the problem into its constituent parts until the range of parameters that define the possible solutions is found [4].

In a morphological system, under normal conditions, function conditions form. Under conditions of imbalance, the constraints of dysmorphic form condition function [5]. Suppose we approach biological processes with linear statistical procedures. In that case, we can establish the importance of parameters by being aware of all the phenomena at work within the system. The underlying network structure decides which entities interact with each other and which remain essential over time. Because of the feature interaction, some system properties may change, sometimes abruptly. The Network analysis of the structural co-dependencies of processes may highlight an *emergent order* of interactions that will itself constitute a form of determinacy involving a direction of system development that occasionally leads to self-enhancement of imbalances more likely. Thus, to properly assess the hidden forces that drive the growth of forms, it is essential to understand the whole structure mapping on which the rules of interaction rest.

The current typical horizon of meaning regarding dentofacial setting is the flow of morphological growth, conditioned by biomechanical rules and properties that change over time. Beyond biomechanical causality, there is a topological causality based on relationships of forms of mutual physiognomy. Class I malocclusions are imprinted on time-eruptive imbalances and local competition for space. Class II and III malocclusions also reproduce traces of a growth imbalance of the bony bases, namely global misplacements. The nodes of the networks displayed by complex biological systems are highly heterogeneous: most nodes connect to a few ones, and a few nodes connect to many others. These different arrangements indicate different types of organisations [6]. As in the facial system, relationships of forms rather than forces govern causality; predictive procedures must favour the analysis of morphological collisions and qualitative constraints they generate during growth. Each facial feature evolves through its connections to other features. It's not just about the direct relationships between bones; the whole network plays a role. Each feature also carries information about its potential variations and what could have been different. Treating one part of the face can affect other parts. Since everything connects in 3D space, changing one feature (like widening the palate) might unexpectedly affect others (like facial height). Therefore, it's crucial to consider all possible side effects before making any changes. The face is a complex system where everything is interconnected, and we must plan the changes carefully to avoid unintended consequences. Problems affecting the three space planes are the most difficult to tackle [7].

While the optimisation pursued by computers takes place through the reiteration of mathematical error compression, biological optimisation coincides with a logic of improving function [8]. Orthodontists navigate a fascinating, intricate world where multiple forces collide. It's not just about following rigid mechanical rules. They must carefully consider pre-established knowledge about facial development, how each patient uniquely adapts to treatment, and the delicate interplay between these factors. This complex dance dictates which facial features become common, rare, or remain hidden, potentially impacting patients' years down the line. Unlike algorithms that deal with cold, complex data, orthodontists recognise the qualitative nature of their work. Raw numbers don't tell the whole story. Data gains meaning only when contextualised within the interconnected web of each patient's unique system. The "quality" of treatment hinges on understanding individual growth potential and the distinct phases of development, making perfect predictions a near impossibility. Clinical experience alone, while valuable, can't capture the whole picture. Furthermore, reducing everything to pure data risks oversimplifying the complex reality of facial systems and growth. Identical outcomes are virtually impossible, as each person's characteristics are unique. While data-driven approaches offer valuable insights, it's crucial to acknowledge their limitations. Ultimately, the key to optimising treatment lies in recognising this intricate interplay between diverse rules, understanding the qualitative aspects of each case, and appreciating the limitations of purely data-driven methods. Only then can orthodontists navigate the complex world of facial systems and achieve the best possible patient outcomes [9].

Orthodontists look beyond the surface. They know facial forms are not static but dynamic expressions of potential and change. Like a seed turning into a plant, each face holds possibilities waiting to unfold. Bone shapes shift subtly, a constant dance of erosion and growth. Every visible feature is a layer built upon the past, shaping the future. So, the orthodontist doesn't just adjust teeth; they guide a living narrative, helping each face become the best version of itself. Forget "brute reality"—patients are journeys in progress. As hidden roads beneath a road travelled, their final form holds echoes of choices not taken. The unseen shapes the seen, with endless possibilities swirling within each face. The orthodontist doesn't invent;

they uncover. They're time travellers who read the potential within and ask "what if" through machines. These machines show unseen pathways, correcting errors and revealing lurking disharmonies. Growth attractors, hidden forces shaping faces, become visible, defying even the wisest doctor's intuition. Treatment must anticipate both the seen and unseen, guiding the flow of change based on hidden connections. Such a vision, not just fixing apparent flaws, is the true art of the orthodontist [10].

Box 3.1—Disciplinary Difference in Modelling Philosophy

Computational models applied to orthodontic discipline try to construct realistic representations of patients, attempting to bridge the gap between data gathering, comprehension, and exploitation. The aim is to make predictions from the sample, i.e., about never-seen future patients. However, some problems still need to be solved in the current state of knowledge. For the reasons we will begin to learn in the following chapters, sometimes training data represent a fragile basis for setting prognostic orthodontic predictions. Systematising the patient by reducing it to numerical formalisms requires a robust conceptual model within which to observe orthodontic facts and data.

When the orthodontist tries to understand what has happened in a patient with a severe malocclusion framework, they must imagine a hypothetical clinical hypothesis that could explain the past, even without genuinely relevant causal details. Having clinical assistance from ideas and tools of statistical physics and computation is an intriguing opportunity for the operator, bearing in mind that finding correlations and recurrent patterns within a database does not automatically mean finding meaningful biological links. Something unexpected, not clinically detectable, must emerge from statistical procedures capable of a multi-sensory reading of orthodontic reality. What appears from these systems is only sometimes interpretable.

Statistical physics and computational intelligence have identified patterns of 'catastrophic events' common to various systems, from financial crises to the breakdown of architectural structures. "Crises" emerge from the co-occurrence of the behaviour of system components and result from repeated nonlinear interactions among constituents. "*... the common denominator of the various examples of crises is that they emerge from a collective process: the repetitive actions of interactive, nonlinear influences of many scales lead to a progressive build-up of large-scale interactions and ultimately to the crisis....*" [11]

Crises express the underlying forces that give rise to 'amplifying critical cascades', often obscured by the dynamics of balancing and self-correction until the moment of their unexpected manifestation. The nucleation of skeletal zones of facial dysmorphism emerges due to correlations of local dynamics triggered by a *collective* process since each event would have a limited impact. The challenge for the orthodontist is to identify the underlying forces of disharmonies before they are clinically palpable. From a certain point onwards, the reciprocal influences underlying the dysmorphism are more robust than the normalisation systems and are disposed of in the facial system following preferential channels.

Current approaches to facial analysis rely heavily on individual data points, missing the bigger picture. This data-driven approach, while applicable, is limited by its focus on what's readily available, potentially ignoring unseen factors like internal balances and long-term growth patterns. Additionally, more data on habits and functions is needed to understand further. Instead of just analysing individual pieces, focusing on how the facial system interacts could lead to more effective treatment plans. Given these premises, we know that "...considering all of the potential sites of error, predicting an exact outcome for a patient using the present system can be quite challenging" [12].

Defining "normal" faces helps identify disharmonies and predict potential patient problems. However, current data often needs to be more accurate and complete and capture the dynamic interactions within the facial system. We need to understand how treatment or even subtle patient differences could change their outcomes. While cephalometric data offers some insights, a more holistic approach is required to consider unseen factors and individual variations for effective treatment. For example, let us consider two male patients, six years old (patient A) and seven years old (patient B), both suffering from Class III malocclusion (see Table 3.1 below) both had a second visit at 9 and 10, respectively. Orthodontic treatment did not start at the time of the first appointment for various reasons unrelated to the operator. These were the initial cephalometric features:

Some negative prognostic signs had made the practitioner suspicious: in our data about a cohort of Class III patients, Classification Trees shows that in this malocclusion, a PPSN angle above 11 degrees often results in unfavourable growth. In addition, an SNA angle of less than 78 degrees indicates a late forward development of the maxilla; recovery is rare in these patients [13]. In contrast, patient B has none of these risk factors. Global skeletal facial divergence is also to the disadvantage of patient A: his SNGoGn angle is 43.8 degrees compared to patient B's 33.8 degrees. Overall, many local and global factors played against patient A. These factors are many local and systemic sites of nucleation of growth distortions. Their interconnection has placed patient A in a high-risk situation of worsening growth.

Predicting future facial development requires a deep understanding of the present. While traditional methods rely on averages and bell curves, this approach overlooks the unique connections network between each facial feature. These connections significantly influence individual growth trajectories, making simple averages misleading. Facial development involves complex interactions between many factors, often leading to diverse outcomes that defy traditional statistical models. We must move beyond averages to predict future facial changes and consider the unique influences shaping everyone's face. Predicting facial development isn't about averages. Individual cases rarely fit generic models. We need a hybrid approach: combining general rules with unique patient information. This plan requires translating individual cases into adaptable regulations while maintaining their specific details. When patterns across cases emerge, new rules form. But these rules still need to hold for everyone, not just a select few. This challenge demands a system that balances general guidelines with individual nuances, ultimately leading to more accurate predictions for every patient. Finally, we get algorithmic information from the behaviours most followed by the patients in the learning set and those who violate the rules because helpful analogies can result from the latter [14]. Patients with extreme characteristics can inform the algorithm about the boundaries of the system's possibilities. It is necessary to have a dataset with many individuals to include rare cases.

Every orthodontic patient is a system which moves in a world of systems. Dealing with the patient's reality means addressing the complexity of the underlying forces that come together at any given moment of growth without looking only at single features or isolated cause-and-effect moments. Algorithms need to categorise the world to describe it better, which entails a certain amount of simplification of the complexity of phenomena; the aim is to define a logic-mathematic circuit capable of linking an input to a response while committing as few errors as possible.

Global models have variables that may be statistically significant for the entire population but may not be for the individual patient. The patient-centred model exploits the data that may be relevant for the individual, not the data useful for the entire population. The aim is to prevent the rules from becoming too complicated, which require complex knowledge structures, such as moving averages and temporal abstraction [15, 16].

A further method to reduce the complexity of analytical procedures is *the Case-based reasoning* approach, which makes it possible to represent a concept (a patient) extensively based on the set of its instances [17]. Predicting facial development isn't just about finding averages. It's about understanding each patient's unique story. Traditional methods rely on general rules, but that misses the mark. Instead, we need a system that learns from individual cases and broader trends. By analysing patients with similar characteristics, we can glean valuable insights not captured by generic rules. This personalised approach allows orthodontists to predict and treat based on unique circumstances, not just textbook examples. Instead of memorising a single map, the orthodontist learns from many different journeys, applying that knowledge to guide each patient's unique path. This way, we can move beyond averages and unlock the true potential of personalised facial development prediction. Case-based reasoning resides in the concept of similarities, the idea that situations recognised as similar in essential characteristics may also be identical in other factors [18].

Predicting development can't rely on simple statistics. We need to learn from similar past cases, not generic rules. This personalised approach helps understand and treat each patient's unique journey. It's about individual stories, not data points. Like other fields like genetics, facial development involves understanding how specific details influence larger systems. We must move beyond averages and see how individual variations "leave their mark" on the bigger picture. Traditional statistics assume probabilistic uncertainty, requiring many similar cases. But not all unknowns are probabilities. Other theories, like possibility theory, could capture unique uncertainties and incorporate expert judgment, leading to more nuanced predictions.

Indeed, a skeletal value outside the norm can be an epiphenomenon of distorting processes initiated elsewhere or even an attempt to renormalise the system, not a causal factor. The growing facial system is an example of a co-evolving system consisting of subsystems in collaboration or competition. We calculate the probability of success of our treatments. Still, we need to be more aware of the underlying causal pathways and main factors that led to the defect, which, if still efficient, could oppose our treatments. We need tools and models capable of simulating the interactions between the elements of the system, which will reveal possible prognostic results after a sufficiently large number of time sequences. The 'hidden half' of the patient's facial characteristics is crucial to the input data to structure a complete diagnostic/prognostic individual pathway [19]. At least initially, we can only offer the machine a picture of patients based on local visible characteristics. However, we must be aware that it is necessary to adjust statistical predictions when there is a high degree of uncertainty in the system and when there are changes in the data [20].

Box 3.2—Systems Uncertainty

The operator's clinical expertise about specific risk factors and the biomechanical notions of force, constraints, and anchorage couples with findings coming from collateral algorithmic concepts about propensity, optimisation, morphological contiguity effects, and growth attractors. Clinical empiricism does not exhaust the reality of the 'organised matter' of the orthodontic domain. Into the spatial and temporal determinations of each orthodontic feature, statistic machines recursively investigate a panoply of collateral aspects. The importance of algorithmic models lies in the possibility of defining different patterns of dentofacial self-organisation as they occur at a given time.

Algorithms should be fed with as many discriminating aspects as possible so that they do not consider all events overlapping and interchangeable, following a patient approximation framework that is too coarse. We must present the raw data to use it as learning data. What we know about the patient's past and what we want to predict must be spelt out as simply and concisely as possible. As stated, identical initial numerical characteristics do not always generate similar sequences of phenomena, which can only be mere analogue correspondences. Indeed, raw cephalometric data similarities, the data we reason about in daily practice, often express partial identities. This reason explains why the operator's judgement is indispensable. The patient dataset must contain at least 30 examples (patients) per characteristic [21]. However, it is not necessarily required to amplify the number of patients disproportionately; the data must be clues to validate or substantiate a hypothesis already in the clinician's head. It is necessary to clarify what level of depth and clarity a feature must possess to convey information about the realistic possibility of probing most aspects of the domain, to define the distance between information and meaning.

The operator must first focus on the correct description level to extract orthodontic knowledge from complex issues. Of some patients repeatedly followed over time, the orthodontist knows the initial facial features and many clinical details: these are the most valuable "labelled" patients, whose outcome of facial growth is known to the operator, from which he can judge whether the development has improved or worsened the initial skeletal morphology. He can use these data to build a function that relates the initial features and the outcome. This data is invaluable for feeding the predictive machine: labelled patients constitute a descriptive framework for learning supervised predictive algorithms, which must transfer what they have learnt in training tagged set to new, unlabelled, never-before-seen patients. Many researchers think the values of the most expressive patients should be represented twice or three times in the learning dataset [22]. The hypothesis is that patients with maximally graphic malocclusion features contain information that minimises the weight of hidden dynamics.

Good orthodontic models don't just show connections; they also predict outcomes for many patients. These predictions could be better, but they provide a range of possibilities with varying likelihoods. However, even the best models can be flawed due to missing data or inaccurate measurements. Jumping to conclusions about cause and effect can be risky. Just because two things happen, one after the other, doesn't mean the first caused the second. Good models rely on data, not assumptions, to avoid this pitfall. Factors influencing outcomes can vary. A change in one measurement might have different effects depending on other measurements, patient age, and individual characteristics. Focusing only on single values can be misleading. Establishing authentic cause-and-effect relationships requires more than just observing things happening in sequence. Without prior knowledge of a link, attending two events happening one after the other tells us they occur together, not that one caused the other. We compute the magnitude of the correlation between two craniofacial variables that move together using the correlation index, which helps us to investigate the variability between the two. The value of the correlation index r varies between 0 (no correlation between two variables) and 1 (the two phenomena follow the same trend). A value above 0.6 indicates a reasonably significant correlation, which means that an increase in one of the two variables correlates with the performance of the other. If it is a negative correlation ("anti-correlation"), one increases when the other decreases. When two variables are strongly correlated but there is no model to explain a causal link, predicting the trend of one from the trend of the other constitutes an inference with a high risk of error. We must identify whether cause-and-effect links are present using probabilistic techniques to find genuine correlations. That is, observing the probability of an effect occurring when a cause is present. There are comical, well-known examples of purely random correlations between phenomena that have no connection, such as the fluctuation of specific stock markets and the consumption of margarine in an African country [23].

Predictive machines must be more comprehensive in identifying recurring patterns in the data and selecting unhelpful details irrelevant to the description or prediction. However, given the vagueness or imprecision of many orthodontic detections, we can get the actual value of many features by observing a phenomenon at various scales [24]. The relevance lies in the structure of the relationships, as we shall see when studying the analytical capabilities of network metrics. Some 'nodes' in the network have far more links with their neighbours than others and occupy a strategic position in controlling many information flows. This additional relational dimension entrusts these nodes with excellent regulatory capabilities over the entire network: the predictive algorithm can observe the collateral capacity of each node.

Instead of focusing on individual tooth shapes, a new approach in orthodontics views misaligned teeth (malocclusions) as part of a more extensive system. This "systems approach" sees the jaw as a connected network where the physical environment and materials are crucial in tooth development and movement. As bones respond to nearby teeth, adjusting tooth patterns during childhood, computer models can help analyse this complex "battle for space" within the jaw. These models, even though based on statistics, can delve into the messy biological reality of tooth development. The goal is to use computers to uncover hidden patterns in facial development data beyond what traditional observations can reveal. This shift from focusing on individual teeth to understanding their interactions within the system, considering the physical environment and materials, opens doors to a more holistic understanding of malocclusions and their development. Practitioners and orthodontists struggle with one of the critical principles of biomedical research: the ineradicable indeterminacy of so many details of biological events. Can we frame the skeletofacial system using a question of angles and segments? How much dispersion procures structurally and functionally uncertain biological factors which disturb our ability to make clinical predictions? Growth accommodates a set of rules, compatibilities, and information that is, as is evident, constantly changing. Do predictive machines let us enter the error of probabilistic judgements that we must continually update as the empirical input data against which to judge changes [25]? Indeed, while we must continuously update data on growth or treatment, the stability over time has provided the best performance of machine learning systems.

Since the obsolete, old "if-then" rule-based expert systems, many decision-support procedures have been developed for specific applications in the orthodontic field [26, 27]. The aim was to reduce the substantial variability of judgement regarding orthodontic decisions (starting with the frequent fundamental question: "to extract or not to extract") and to diminish the subjectivity of many choices that arise in the daily life of the orthodontist. A variety of application topics covered are treatment planning, the decision to proceed with orthognathic surgery, the prediction of the onset of temporomandibular osteoarthritis, the risk of jaw canine impaction, the best type of headgear; the assessment of respiratory dynamics; the risk of obstructive sleep apnoea; skeletal bone age assessment, and other [28]. The accurate localisation of cephalometric landmarks using computational tools has led to a mitigation of the interpersonal variation in landmark tracing and related errors in diagnosis, with 5–15% better accuracy with machine learning of landmark detection than traditional methods [26, 29].

3.1 The Science of Combined Effects

Observing the elegant symmetry of structures in matter (think of a snowflake), we tend to imagine universal, natural, unifying laws and obligatory paths linked to dominant evolutionary force fields. In the enormous variety of material structures, we suspect each must have its law of formation and a specific possible framing system. The general laws of the nucleation of facial dysmorphosis must match the specificity of the individual organisms. In Chap. 4, we will see that the laws of Synergetic, the *science of combined effects*, advise considering the many parts of a living system, as the whole system dictates the condition for understanding. The German theoretical physicist Herman Haken hypothesised unifying Synergetic laws of general validity that embrace chemistry, physics, biology, and economics, which concern the emergence of structures in which we see a spontaneous inherent ordering dynamic. In the self-organisation of these structures, Haken found a global 'authoriser' effect produced by the coordinated dynamic of the various parts, which governs the behaviour of the parts and 'enslaves' them. The system's authoriser is in its structure, which allows one to imagine the future from the past. In the formation of structures, the shape of the contour and the macroscopic morphological coherence of the system (which must also be present in the micro-components) are of help [30]. This approach is similar to the binomial potency act that we have seen in Aristotle's Metaphysics. We will also know that it is not far from the statements of Oriental medicine about generative dynamics of health and illness: the *propensity-unfolding* paradigm.

Following the strategies of Machine learning procedures, favourable facial growth factors from which to profit can be sought in the data, not in the rules. In Chap. 5, we will see that clarifying the "potential of the situation" means exploiting the favourable therapeutic factors to learn: "...all of reality is, in fact, nothing but a succession of transformations. The action is salient, effective, and flashy, but only the transformation is effective. The action, at best, can represent a focus-clot-contraction in the continuous course of things" [31].

When you know the system's structure, you can check how much of the variability in the data is in the model. The tolerability of facial dysmorphisms depends not only on the doctor's judgment about aesthetics but, above all, on functional/somatic flexibility of the overall facial system, on the mutual accommodation between skeletal muscles, joints, fibrous, and connective compartments. In this accommodation, new information is created and disseminated [32]. Morphological homeostatic regulation during morphogenesis is modelled based on the mutual positional information of the various local parameters involved in implementing the body plan. The implementation conducted merely on local structures alters the geometric properties of the organ, affecting the global plan. The local process required to create the structures necessary for growth depends on the pre-existing whole-body plan. The transformation generated by growth unfolding occurs globally rather than locally: the whole complex transforms. On the other hand, morphological whole homeostasis limits the extent to which genetic variability passes into phenotypic variability during morphogenesis [33].

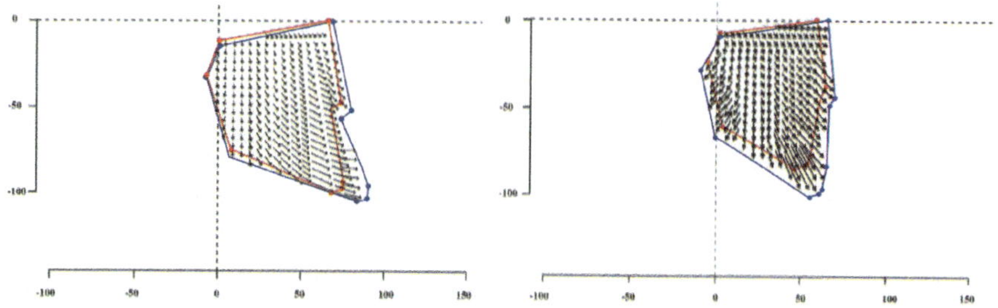

Fig. 3.1 Forward growth (left) vs downward growth (right) of two class III patients (7–13 yrs)

The craniofacial system is not an isotropic body that deforms equally in every direction in response to structural pressures (growth vectors- see Fig. 3.1).

While computer predictions for facial growth hold promise, they still wrestle with some limitations. Predicting too far into the future in such a complex system introduces more uncertainty. Similar to weather forecasts, shorter timeframes, around three years, in this case, offer more reliable predictions. Additionally, grouping patients by age and gender further enhances the accuracy of these estimates. It's crucial to remember that facial development isn't a linear process. It's a dynamic interplay between biological factors like growth hormones and the physical constraints imposed by existing bone structure. Each element in this system influences and is influenced by others, creating a complex dance that unfolds over time. Acknowledging these intricacies, we can utilise computer predictions more effectively to understand and anticipate facial growth patterns. Despite the non-linearity of facial growth, the predictive model uses linear regression to define the importance of individual factors for the total number of patients, exploiting subspaces of patients with similar characteristics (classified by cluster analysis in Chap. 5). The *personalised* model constitutes the final goal of orthodontic computational prediction. Depending on the case, we can use algorithms to implement personalised, predictive models (possible additional specifications about this topic results about LIME algorithm, in Chap. 5, and *Transductive* learning, in Chap. 4 [34]. Finally, a fuzzy prognostic procedure will have to be employed, resorting to the laws of *fuzzy logic* (debated in Chap. 5). In short, computational procedures exploit the capabilities to decipher inherently complex aspects that are not clinically palpable in many ways, including the whole "system's connecting structure".

In several research domains, the information explosion linked to computers has enormously increased data flow, enabling the algorithmic formalisation of disparate phenomena. In the orthodontic field, a wealth of empirical documentation now supports a plurality of models of facial development. Luckily for orthodontic practitioners, avoiding the dispersion of concepts and models. Machine learning algorithms still have a chance to continuously improve, monitoring the facial system as it grows, choosing the most relevant units of analysis each time, and exploiting not only the information from orthodontic signals but also from the noise resulting from the uncertainty in the data. As we shall see, is the operator's proper understanding of the patient's facial ecology defining the most appropriate investigative procedures and the significant results? (Box 3.4).

Box 3.3—Facial Development

Facial development involves a dynamic interaction between multiple variables. A clinical sign can only be meaningful because many factors converge on it. Dynamic interactions between variables that each patient brings form mutual causal relationships of constraints. Consequently, a facial feature outside the normal range only becomes meaningful at a particular stage of development and if it can rely on cooperating or competing contiguous structures.

During the facial morphogenetic process, the laws of self-organisation obscure the underlying forces that govern the appearance of dysmorphosis. A qualitative factor connects to the appearance of new 'structures' and 'forms', which determines and delimits concepts of normality, acceptability, disharmony, and malformation. Concepts originate from morphological and radiological findings, in the impossibility of appreciating in the individual patient the part attributable in detail to genetics, developmental noise, co-occurrence of competition phenomena (oral respiration and deep bite) or collaboration phenomena (atypical swallowing and open bite). Whether the observed phenomena had a reiteration over a sufficiently large number of time sequences, they can create a distortion zone. Every growth path can end differently because of a trifle, deceiving our desire to derive general laws valid for every patient.

While computer predictions can potentially understand facial growth, they have limitations. Just like weather forecasts, predicting far into the future is unreliable in this complex system. Shorter timeframes (around three years) and grouping patients by age and gender increase the accuracy of such predictions. Remember, facial development isn't a simple progression. It's a dynamic interplay between factors like growth hormones and existing bone structure, where everything influences everything else. By acknowledging this complexity, we can use computer predictions more effectively to understand and anticipate facial growth patterns, but with realistic expectations about their limitations. As in all medical domains, in orthodontics, 'normal' falls within a range of values rather than a single average value [35]. It is not trivial to define the acceptability of a pattern, to divide what is visible from a web of previous independent biomechanical events, or from the succession of interacting organisational dynamics. What may be highly significant within a subgroup of patients may not be so in patients with other characteristics. Although the discussion between average and ideal values is outside the scope of this book, we note here that the complexity of the subject and the questionability of the diagnostic criteria have led to the assumption that only 9% of the population of young Americans aged 12–17 have a normal occlusion [36, 37].

Figure 3.2 shows the contribution of the variables to define the maxillo-mandibular imbalance in a group of Class III subjects (6–17 years of age) subdivided into two groups with mild and severe maxillo-mandibular imbalance. We see a clear difference between groups.

In our series of 144 Class III patients followed over time (male and female, 6–15 years of age) who did not undergo treatment for various reasons, 27 had an initial Wits appraisal lower than −6. Re-tested after two years or more, all 27 had worsening Wits scores. These scores lead us to believe that a sagittally unbalanced skeletal structure constitutes an unfavourable constraint for subsequent facial growth and that a propensity with such characteristics can influence the global unfolding of the system.

The concept of normality in orthodontics links to morphology, dentoskeletal relationships, and the time in which the interactions occur. The *temporal abstraction* algorithm can integrate various data types and extract developmental characteristics from the patient's history by having multiple temporal records of the same patient.

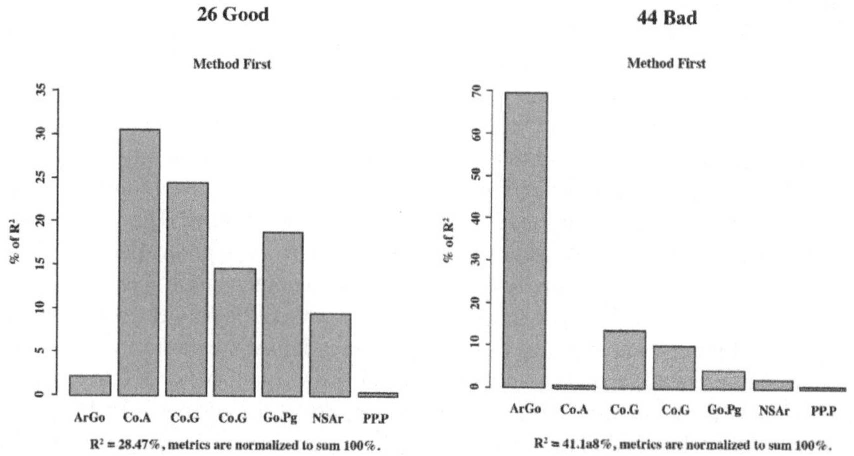

Fig. 3.2 Relaimpo "First" Analysis: Out of 70 patients with Class III malocclusion followed over time (male and female, 7–16 years of age), 44 increased maxillomandibular sagittal skeletal imbalance (worsening of CoGn-CoA imbalance, referred to Class III malocclusion CoGn-CoA mean values for age and gender) throughout growth ("Bad", right), while 26 maintained stable imbalances, or improved the dentofacial sagittal relationships ("Good", left). Relaimpo "First" analysis showed a homogeneous contribution to growth by various cephalometric features in patients with satisfactory skeletal growth. At the same time, the ArgoMe (Gonial angle) appears to have contributed more than all other features in the group of patients who exhibited progressive worsening of skeletal sagittal relationships. Relaimpo allowed accurate variance partitioning and accurate contribution each predictor makes towards the dependent variable (in this case, a good-bad skeletal growth), even among correlated predictors. Relaimpo analysis can also distinguish the contribution to an imbalance resulting from the action of individual characteristics on the outcome (First method) from the contribution to an imbalance resulting from the interconnection of several system characteristics (LMG method). The LMG method provides further information on the components of the interactive organisational structure of facial imbalance. Some authors believe that, beyond the known factors, "developmental noise" is responsible for 40–70% of system variability. We will see in the following that there is the possibility of defining the role played by 'latent variables' [8, 38]

We know the true tell-tale of the patient from his past time trends, from which the prediction for the future can be derived [39, 40].

Reasoning a posteriori allows one to field logic modules related to changes in the patient's facial structure over time: initial diagnosis and related assumptions must comply with the new findings. Notions such as the persistence of certain features, global skeletal propensity, recurrence of patterns, nucleation of anomalies, and co-dependencies between data emerge by proper evaluation over a given period. The change in structure over time provides adequate support for prognostic reasoning, as no morphological data is timeless in orthodontics. The patient's past trend guides the visualisation of his future values because "the past is the prologue of the present" [41]. Operational knowledge about the patient must be based on the correct level of abstraction regarding the summary of the past and present. Temporal abstraction transforms time-stamp points into an interval-based representation of patient data. The importance of the individual variables emerges from their temporal sequences. The presence of an anomaly may not be significant as such, but it can be in the context of a specific group.

Computers can help understand facial growth by focusing on critical details instead of everything. They also consider how long each factor influences others, essential for predicting future development. Additionally, each feature links to various aspects like "belonging to a certain group" or being affected by other structures. However, computer predictions based on past data also have limitations. Our initial assumptions about "normal" might lead to biased results. The orthodontist must know these biases and factor them into their knowledge and experience. This approach includes considering hidden factors the computer might have identified, like specific influences between facial features or competition between different jaw areas. Combining computer analysis with the orthodontist's expertise can provide a more comprehensive understanding of facial growth and better predictions for individual patients. However, we are still determining if the current snapshot of the patient will still be valid in the residual growth. It seems particularly appropriate for the orthodontic discipline what Michael Blastland wrote in the seminal book "The Hidden Half": "It is a world where the search for regularity, general laws and causal mechanisms must continue, always, in a state of perpetual uncertainty in which it is simply impossible for us to know what we might have missed … the problem is the myriad details, irregularities and intangible differences, complications, and unintended consequences. It is a situation where something applies here but not there or now but not then. The quest for knowledge, in practice, is an endless struggle between data, theories and hypotheses in search of a replicable mechanism…" [19].

The prediction regarding malocclusion assumes that no underlying linear laws and constant biomechanical forces are acting on the development and that one can only reason based on what one sees in the patient. However, there are underlying quite reliable trends. The presence of full tooth gear "is an assurance of stability over time, in terms of tooth alignment". Indeed, the totality of patients who initially presented with a Class I molar relationship still had this occlusal pattern 40 years later. It's a good starting point if we have confidence in what F. McDonald and A.J. Ireland reported: "…although there are no reliable mechanisms to predict the reliable potential of an individual, it is usually safe to assume an average degree of facial growth. Therefore, the relationship between the maxilla and the mandible will change little with time" [42].

In 2011, Piotr Fudalej and co-workers conducted a systematic survey of the orthodontic literature to determine the most reliable indicator in predicting unsuccessful treatment outcomes for patients with Class III malocclusion. As many as 232 publications on the subject were examined. Despite meticulous work of selection and exclusion towards reports judged to be of low quality, the studies reported different parameters with the best predictive ability, from the Gonial angle to total mandibular length, from Wits appraisal to branch length and Overjet. In the end, the various studies indicated 38 different predictors! [43]. There are forces inherent in the growth of the structure, which powerfully exerts a constraint on the individual variables, depending on the number of links each component has with the other nodes of the network. There is a considerable variation of mutual feature interconnection over time and a possible enormous variation in the final effects of growth. The overall morphological structure influences the impact of one variable on another specific variable (we saw it with Bayesian networks). Indeed, in many biological-type organisations, the functioning of each component is, at least in part, controlled by the general state of the system [41].

The properties of the different facial skeletal zones must conform to the internal coherence of the organ. In the network of reciprocal relationships, each variable contains a share of the Influence of other variables, and the value of each variable depends on the value of all the others. The dentofacial structure is more than an assemblage of features. Still, as a dynamic network of interconnecting processes beyond the visible morphology of defects, the orthodontist must observe the strength of the interconnections that determine the nucleation of defects and their evolution over time. The findings of Fudalej's work show that no single component can determine the success of treatment: the growth dynamics devoted to establishing the global coherence of reciprocal connections can decree the structure's balance or imbalance. These may or may not be overwhelmed by the biomechanical forces of treatment. No single 'order-giver' (as in Haken's philosophy) can 'command' the success or failure of a treatment. No single component possesses this fundamental capability.

In particle physics, many current Western schools of thought refer to the internal coherence of natural systems as an expression of reciprocal relationships of components. Nature not only shows atoms or elementary particles but also interacting 'self-consistency'. Each component shows its coherence with itself and reciprocally with the other elements. As Denis Noble wrote, higher-level properties also cause lower-level behaviour involving interactions. The genome does not automatically define the organism; there is no privileged scale on which biological functions appear, as "…higher levels in biological systems exert their Influence over the lower levels. Each level provides the boundary conditions under which the processes at lower levels operate. Without boundary conditions, biological functions would not exist…" [44].

In his fascinating book 'The Tao of Physics', Fritjof Capra narrates the journey that led Geoffrey Chen to theorise the foundations of the *bootstrap* hypothesis. The bootstrap is any entity that can stand on its internal consistency. This internal coherence represents the essence of all laws of nature, the organising principle of the interactive network to which the components must conform due to their existence as parts of the whole. In this shaping reality, the organising principle lies in the mutual coherence of the elements, as each entity depends on all the others and is a consequence of all the others [45]. Even in Eastern philosophies, the explanation of nature lies precisely in showing its unity. There is a network of reciprocal relationships where all events interact "...in such a way that each of them contains in itself all the others" [46].

The primary work of the orthodontist is to seek a reproducible rule that gives value to past observations in predicting future events. He must research the process that led to the visible morphology by observing not so much the general law and the reproducibility of the measurements but rather the specificity of the reasoning of the individual patient. In formal terms, he is looking for historical data vectors applicable to the whole problem space. The inductive model starts from past patient observations, which help decipher the new data vector (the new patient). A general rule from the data works in at most 70% of cases [7].

Compared to inductive reasoning, the approach based on *transductive* inference creates a model based only on a specific set of data vectors, not on the whole problem space. A sub-model of data more in line with the patient's characteristics under investigation is necessary: for example, we must cross-reference a patient only with patients who are similar in age, gender, and morphological characteristics. Optimisation of the variables and appropriate characterisation of the background of each patient facilitates the creation of a personalised model. This approach seems particularly useful for extracting the crucial principles underlying specific facial dysmorphic situations.

Individual variables express their value to the fullest when compared with a background of consistent variables from large numbers of patients with similar characteristics. Certain assumptions are necessary for unlabelled data to help define the boundaries of the problem. First, following the cluster assumption, it is likely that the closest data have similar characteristics. For example, similar facial features = higher probability of similar growth. We have seen that this assumption is the logical-statistical basis of Case-based reasoning. Data in the same morphological cluster or belonging to patients with similar characteristics in various ways share features relevant to further dimensions, e.g., temporal. Transduction aims to combine the best labelled and unlabelled data to gain new knowledge and benefit from this combination. The transduction uses unlabelled data to modify hypotheses obtained from labelled data alone. The basic concept is that each patient possesses personal characteristics and characteristics of their group, so it is helpful to understand the environmental characteristics well.

The transduction inference seems more appropriate in the medical field and even more so in orthodontics, where the problem is not the model but the individual condition of the patient, his environment and specific history [47, 48]. In addition to patients whose history is known, the algorithm can access other data whose labels it does not know. From this unlabelled data, we can extract further information. Transduction inference manages to reason from the particular to the general, based on the boundaries of possibility observed in unlabelled subjects. In other words, information flows from labelled patients (whose outcome of response to treatment is known) to unlabelled patients (to predict whether they will respond successfully to treatment). Start with only labelled data and recursively add some unlabelled elements to the decision function until we attain a final stable state. Whereas inductive inference attempts to estimate a function throughout its domain of validity, transduction stops at the particular of each patient, avoiding arriving at general rules (which, as we have seen in orthodontics, have many exceptions). The advantage is to use all patients, not just supervised patients, to find interesting structures in the data and to be able to make predictions with little data. All in the hope of decreasing the uncertainty of what will happen.

Systematic extensive clinical and radiological analyses, although accurate and conducted using various sensors scattered in multiple areas of the facial system, uncover only some of the factors involved in the patient's orthodontic situation. There is a parallel functional/morphological reality that we do not see hidden in the noise of variables. We risk jumping to prognostic conclusions too quickly when we know less than we think.

3.2 Make Cephalometric Quantities Mean More

Following the computational logic, the orthodontist can discover alternative developmental histories. The facts reconstructed through this lens allow us to appreciate the location of unsuspected dysmorphic elements and possible spill-over to other components, near and far. It is a preliminary reconstruction: proceeding with successive analyses, the system changes its appearance, and the complexity increases in parallel with visualising the variety of interconnections. The practitioner can refer to various geometric reference systems when choosing between different diagnostic methods of reading cephalometric data. However, each piece of cephalometric data provides only a tiny window into a limited aspect of reality. Cephalometry represents an attempt to stiffen a fluid and reactive matter. In this way, we can locate local aspects favouring the worsening of facial dysmorphia. Still, above all, it is necessary from the outset to identify specific *circuits of accumulation* and *disposal* of global skeletal imbalance. The aggregation of unbalanced factors, which resulted in an appreciable defect, represents the memory of how things unfolded during facial development. It testifies to the extent of the distorting processes, the ineffectiveness of the disposal processes, and the uniform distribution of defects in various system points. The anamnestic reconstruction of the patient's morphology aims to bring out deep structures based on aspects often hidden precisely because of their complexity. Identifying imbalance patterns makes it possible to search for structural relationships between morphologies and influences between facial zones and relate different parts to the same morphology. The facial system establishes patterns to optimise skills for planning one's development: when dysmorphic and renormalisation dynamics place excessive or contradictory biomechanical demands on craniofacial adaptation, stress faults can emerge, along with bottlenecks, deformations, root apexes-basal bone constrictions, shape changes. Harmful interactions introduce leaks into the system in the form of micro anomalies, which cause the system's energy to decay along with the onset of painful symptoms (the more complex a global dysmorphia is, the more difficult it is to counterbalance it by dentoalveolar renormalisation processes aimed at maintaining cusps-fossae contacts).

Some prepubertal local characteristics may have little to do with the specific properties of the final facial components. For this reason, it is necessary to think in terms of large-scale structures, deriving from processes of collective organisation rather than on single early variables. The premise is that the operator can interpret the algorithm's response, considering possible "emergent properties" and providing the machine with an adequate conceptual map of patient age, connections, and context.

As relationships among features are the core of dentofacial reality, proper language is needed to interpret them. The concept that the appreciation of individual parts is not enough to understand the behaviour of the whole must be clear to the orthodontist. Because 'the overall system is more than the sum of its components', none of the components can tell the whole story. Several local features become more evident in the light of the clarification of the entire context, so the global and regional levels are separate. During morphogenesis, billions of cells demonstrate a supramolecular coherence, manifesting only one morphology amongst the many potential ones.

In biological networks, every element interacts with every other, and each component's decision depends on what everyone else is doing. The emergent phenomena typical of complex systems consist of the transition from the components to the creation of totality: the laws of autopoiesis teach that A produces B, which in turn makes A. It is a circular system in which each component tries to realise its potentialities in accordance (sometimes in disagreement) with the overall structure [49]. The overall facial structure allows the most effective couplings and interactions between the skeletal parts. Thus, the global structure constrains the local zones for the growth constraints of patients with severe maxillo-mandibular imbalances. Still, the modality of local solid interactions is responsible for the emergence of the totality. In medical phenomena, researchers often use linear regression techniques to look for prediction, regularity, and reproducibility to find a logical order in the succession of facts, while in nature, the rule is disorder and irregularity. The heterogeneity of facial forms teaches that the progression of malocclusion can differ from person to person, even starting from similar characteristics. This heterogeneity contradicts the cluster assumption (components of a cluster tend to assume the same behaviour over time): apparently, a predictive statistical procedure adapted to orthodontics needs further specification.

Nature does not produce the regular shapes of Euclidean geometry, such as spheres and triangles: lightning does not propagate in a straight line, the islands' coastlines are jagged, and the branches of trees are not uniform. There is also a temporal hierarchy in the planes of reality of living matter. Some periods are more favourable than others for growth and corrective action. According to Thowald Dethlefsen, to understand the meaning of what surrounds us, we should not divide and make comparisons into horizontal planes of reality but proceed by vertical subdivisions by *chains of principles*. A hierarchy where every interaction and mutual Influence become visible, and they acquire significance in the coherent plane of existence [50]. Using a matrix of thinking at various planes of reality allows one to appreciate the vertical correspondences and propensities that dominate organogenesis. Traditional acupuncture philosophy discovered the whole in the hand, nose, and every cell. In Oriental medicine, there has always been the assumption that each single part contains the totality of the organism, so each part includes information on the whole.

Cephalometry shows the inclination of the teeth and the position of the bone bases concerning the teeth and facial skeleton. As is apparent, in cephalometric geometry, we cannot hope to find the effect of interconnections, dominations, gregarious behaviour, mutual compensations between features, etc. However, it may be helpful to remember that each facial characteristic relates to a specific subset of connected attributes. Network theory explains that relationships between structures affect system components. Click or tap here to enter text' behaviour and that every feature can be a point of instability or strength. It describes how vulnerable facial features can become the first adopters of dysmorphia.

At the end of the last century, Western researchers have used the General Systems Theory [51] and the formal relationships of the discipline of Mereology in disparate fields to explore the relations that bind the parts to the totality. These efforts were motivated by one desire: to bring out deep structures based on processes often hidden due to their complexity, using chains of mathematical principles of hierarchy that could link the interconnections between the various planes of reality [52].

Before defining a future geometry of form, we need to know the basis of the current structure and what is possible and not in the future, given the ensemble of predictable interacting micro-phenomena. The potential and the virtual in biological systems are more prosperous than the natural reality, which realises only a few morphological potentialities, deformations, teratologies, or shape changes. Following the responses of the digital explorers, the orthodontist can define the most probable accommodation of the morphogenetic field, which organisational principles and final structures are more likely, given the premises. He must explain how characterised the malocclusion is in each patient, for example, to what extent a patient suffering from a Class II malocclusion with hyper-divergence belongs to the cluster of subjects of the same age, gender, malocclusion, and divergence. High cluster membership means that the features of that subject's facial disharmony are highly expressed. It is a critical survey whose implications emerge from, among others, fuzzy clustering methods and Sankey diagrams (see below). It leads us to hypothesise why the visible morphological structure if equipped with specific crucial stabilising characteristics, will hardly change during future growth. We will see that computational systems offer many ways to visualise whether strongly constraining morphological features are present. Electronic decision support systems can highlight specific hierarchical chains of information structures that link different facial morphologies. In principle, a local severe morphological distortion prevailed over attempts at renormalisation; the whole system struggled to maintain a specific structure and a certain rate of development and change. Thus, the results from these analyses may be necessary for most patients but only for some.

The surrounding skeletal and alveolar zones of a dysmorphism change to encircle the defective area. In this way, the struggle of local morphology leads to a change in the global context. When dysmorphia tends to overwhelm surrounding structures and overcome renormalisation systems, the entire system loses flexibility, forced to an ever-stronger amplification of the compensatory adaptation. The spiral of morphological matching between the facial parts triggers a dynamic of mutual accommodation, sometimes in a morphological configuration that constitutes the most immediate way of dissipating the imbalance. The new matching causes an increase in the exchange of information, coming from the facial system itself rather than from the external environment, following preferential dispersion circuits and the uniform distribution of maximal stresses. (according to the "Constructal law") [53]. The distribution of imperfections originates in the system's geometry (shape, structure). Only by learning both the contextual inner organisation of interaction structures (which must persist for a specific time) and the logical path that motivated the visible state is it possible to make probabilistic assumptions about the future spatial organisation. These dynamics are highlighted by studying the spontaneous aggregation of features, visualised using the analysis of Network stress, shortest paths, and motifs (see Chap. 6) [54] (Fig. 3.3).

Following concepts rooted in the biological world, these predictions refer to the formal balance of geometric relationships versus mere clinical and radiological observations [55]. Gregory Bateson emphasised the subtle geometric-clinical matching with the following example: for a living being, the difference between the four sides

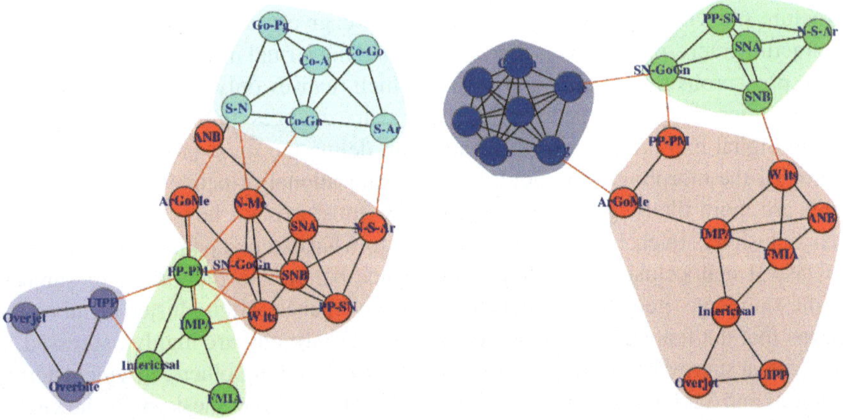

Fig. 3.3 Modules (motifs, communities) [54] of cephalometric features of 35 Class III subjects incurred in unsuccessful orthodontic treatment, with no overjet correction (left). On the right are 35 subjects of the same age with successful treatment. Communities on the right are more transparent (less bridges between the communities). Modules are potentially valuable tools to predict unknown interactions involving a few nodes. These organised interactions allow higher-order functions such as amplification; they favour plasticity and homeostasis and represent the functional capability within the network. A strong correlation between attributes gives rise to a new attribute

of a square and the three sides of a triangle is not at all 'four minus three, which makes one' (the simple mathematical solution), but rather a very elaborate difference in configuration and symmetry' [56]. Machines can develop a different way of seeing the world compared to doctors. They can take a large amount of data into account simultaneously, differentiating the signal from the noise and establishing inference networks between input and output regarding probability and correlation (not certainty). In clinical orthodontics, the morphological imbalance holds past and future events together. As observed by LeCun and co-workers, if a convergence of defects (a modulus) appears in one part of an image, it can appear everywhere because the parts share the same pattern in every part of the system [57].

The weight assigned to each numerical craniofacial feature must contain information on how much each variable managed to outweigh the antagonists or the co-operators during the development process. We need to enter each value together with the related values for the same characteristic (for example, the angles which express the divergence between the maxilla and mandible). The orthodontist needs a conceptual map to refer each value to the internal coherence of the facial system, whose properties of each part depend on all the others. A fuzzy theory is needed to minimise the friction between biological reality and the multiple interpretations of reality derived from computational models. It is necessary to imagine the presence of 'organismic constraints' to limit models that are too abstract, based on virtual (non-biological) hypotheses, or models based on too many pre-empirical intuitions.

The meaning of each feature is not explicitly present in the input. Still, we can expressly consider and add to the raw features (1) the importance of the skeletal counterparts, (2) the strength of the colliding components (usually present in the same module of the network), (3) the residual growth of each patient, (4) the deviation from the mean of each feature, (5) the overall morphological internal coherence (i.e., the propensity), (6) the local micro-constraints acting on growth.

Integrating the dimensions of craniometric variables using computer systems can establish a new relational structure among features. Spatial relationships between counterparts bind disparate independent characteristics. For instance, the importance of CoGn mandibular value increases by coupling with a geometrically maxillary counterpart variable CoA and with different pressures coming from the global homeostasis of the system. Following physicist Dendre Gentner, there is a "structural predicate" that influences each variable: the orthodontist must offer to the machine a variety of inputs endowed with the complex articulation of orthodontic data, hoping to gain insight into the essence of the growth of the facial system [58].

According to Donald Enlow, distinct regional modifications in the developmental phenomena co-occur between structural parts and counterparts. Balancing these parts ensures harmonious growth [59]. Auto-compensation is a development feature: it causes a specific imbalance in some areas to compensate for the effects of disproportions in other regions to maintain a structural balance. Each craniofacial area is related to structural and geometric counterparts of the skull and face. If each party and its counterpart expand to the same extent, balanced mutual growth will result. Differences in the respective amounts or directions of growth also produce imbalances. Tuberosity stimulates the horizontal development of the jawbone, which grows backwards by moving the entire maxilla forward. The counterparts of the jaw are the upper cranial fossa, the nose-maxillary complex, the palate, and the body of the mandible. However, it is not a matter of a binary comparison between the size and shape of individual bone morphologies but a comparison of several interrelated counterparts made of a network of components and processes. Sometimes, the excessive branching of attributes hinders the depth of developmental analysis. The hope is that complex articulations of characteristics can contain aggregate information formalised via algorithms and constitute a generalisable orthodontic rule.

The progression of multiple sites of local dysmorphoses tends to affect the closest nodes in the network, thus creating an interactive gregarious behaviour (following a 'contagion pattern'). "In physical terms, the macro-state of a system must reflect its micro-state (as the global temperature and pressure of a container of gas reflect the velocity and collisions of individual molecules)" [60]. The need is, therefore, to unearth the underlying dynamics and unravel the mirroring between local micro-states and the macrostate of the system in the simplest way possible. "...An important point in statistical physics is that different microstates can correspond to indistinguishable macrostates. For example, various possible distributions of molecular velocities can produce identical average velocities. [...]......even if [statistical mechanics] makes statistical predictions, in many cases most possible microstates produce similar macrostates, so the prediction of that macrostate has a high probability of being exact" [60].

The assemblages coming from different computational approaches can have a multiplier effect in interpreting the behaviour of the facial system, i.e., the structures based on the rules of interaction. Every facial morphology contains the sediment of past processes, which can only have a coarse set of approximations. But it is not just the effect of the inevitable measurement error. When paths result from the interconnection of other paths, statistics pile up data under the assumption of common patient similarity. This assumption could be more robust in orthodontics, where we do not have many patients to hide eccentric ones. Structure progressions tend to be distorted by effects common to complex systems, such as "preferential attachment", "percolation", and "fields of attraction" [61]. These effects can result in procedures capable of framing each patient as the centre of a network of relationships, past or future, actual or possible, of related evolutionary strategies and defects spreading. In this framing, the inclusive character of a grouping of local reciprocal micro-dynamics becomes a form of determinacy. The new orthodontic scenario emerging from the network of relationships disclosed by algorithms can separate fundamental characteristics from the noise of incidental random features, improving how the orthodontist's intuition results in a deeper understanding of the patient.

The questions the operator puts to the machine may, therefore, sound ambitious yet sustainable: ".... what can I deduce about the future behaviour of the facial system from the defects I see?" ".... what kind of local prints will a global skeletal structure leave in the future of the patient?" "... localised defects that I see will be improved or worsened over time by the structural pressures from overall skeletal morphology?".

Despite the premises, a complex system is not necessarily a complicated system. It has only three essential constituents: the building blocks, the interconnection flows, and the objectives. The teeth, the basal and alveolar bone, soft tissue, and the temporomandibular joint are the first components in the complex dentofacial system. The second components are proprioceptive information linking components, self-correction dynamics, and adaptive phenomena. The third component is the system's whole *propensity*, resulting from the many heterogeneous parts to preserve and optimise the internal order. System propensity is hypothetical in Nature and requires adherence to an organicist orthodontic philosophy. It is easier to study the individual morphological elements rather than the network of interconnections. But morphology cannot tell everything about the pervasive phenomenon of mutation nor reduce it to trivial connections of action and reaction: it is crucial to study the interconnections because if they change, the system's evolution over time can vary significantly.

3.3 Accumulation of Causes

During interplay among many dynamic forces operating in the evolutionary facial system, there may be imbalance dynamics due to the predominance of one or a few local factors contributing to a malocclusion, visualised by simple, linear computational procedures (Fig. 3.1). Relaimpo analysis uses multiple regression procedures better to understand the role of each predictor in the regression.

According to the systems thinking proposed by Donella Meadows, for the same dysmorphosis to allow a prediction, it is crucial to focus on the extent to which the constituent elements are stored and disposed of overtime over the whole system. The constituent elements in different systems can be material quantities such as liquids in a tank, money in a bank account, or information in a financial network. Causal, adaptive, and retroactive connections between components are difficult to detect; we must study them to understand and influence them.

Systemic "things" increase and decrease in the system, depending on the magnitude of the inflows and outflows, just as would happen in a bathtub where we could change the faucet flow and the behaviour of the outflow plug, as shown in the figure above. In the bathtub metaphor, the water that accumulates over time is referred to by Meadows as the 'stock', while the circuits inflow and outflow of the liquid from the bathtub are the 'flow'. In orthodontics, the metaphor of stock and disposal corresponds to the amount of dentofacial imbalance that, under certain circumstances, is generated and increased during growth. The matter becomes complicated if we imagine that different reinforcing circuits at work tend to increase the liquid flow into the tank, connected with feedback-balancing circuits that maintain the balance of the amount of liquid. We will see the water level fluctuate depending on the prevalence of one or the other, i.e., the change in dominance between feedback circuits. Generally, since balancing circuits have more inertia than reinforcing circuits, renormalisation processes will be slower, so there may be fluctuations in the outcome of the process. Moreover, as a natural biological system exhibits simultaneously different reinforcing and balancing systems, the overall effect of competition will produce a global dominance effect that characterises the system's global behaviour over time.

Indeed, uncertainty concerns the structure, i.e., the network of connections within the system, and this uncertainty can drive us to error.

The stock represents the visible extent of the accumulation of the defect. At the same time, flux defines the balance between the number of defects formed and those the system disposes of over time. In a nonlinear system, changes in one variable can act on another and accelerate it enormously. Suppose my procedure caused a 5-degree increase in the maxillo-mandibular divergence angle. In that case, I expect a much more significant facial imbalance effect in a patient with a starting value of 31 degrees than in a patient with a value of 24 degrees. In a system that is not linear but exponential, where there is a combination of factors acting simultaneously, we must try to *work by subtraction.* Instead of adding devices that risk increasing the effects of non-linearity, try to limit the stock by stemming the flow of defect accumulation. Subtracting the flow of oral respiration from the system can avoid the application of corrective devices for hyperdivergence (which are often disappointing).

During morphogenesis, the design generation of forms is a physical phenomenon involving the uniform distribution and disposal of imperfections ('Constructal law') [53]. The distribution of imperfections generates the geometry (the shape, the structure) of living systems and facilitates the body's flow of stresses. The treatment forces applied at one point of the skeleton are distributed over the entire organ through *preferential stress circuits,* referred to in network theory as *shortest paths* or *stress.* More robust bone materials form where stresses are higher to facilitate the dispersion of forces (as in the zygomatic abutment above the first molars of the maxilla) (see Fig. 3.4). These cooperative solutions correspond to assets of organisation steps occurring over the whole system. We indicate them (with different shades of meaning) as "paths", "channelling", "propensities", "morphological integration", and "amplifying critical cascades".

As the number and the expansion of defects increase, exceeding the homeostatic capabilities of the system, a new "phase" appears, leading to a global skeletal disharmonic structure, which we interpret as a *percolation* process. With network analysis,

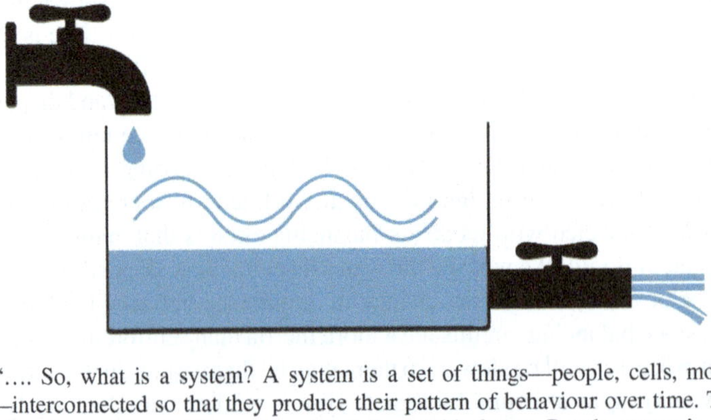

Fig. 3.4 ".... So, what is a system? A system is a set of things—people, cells, molecules, or whatever—interconnected so that they produce their pattern of behaviour over time. The system may be buffered, constricted, triggered, or driven by outside forces. But the system's response to these forces is characteristic of itself, and that response is seldom simple in the real word..." [62]

we describe the laws that govern the almost sudden nucleation of defects in limited lean zones and the formation of stress basins that propagate to surrounding areas. Indeed, network analysis grasps the subtle reciprocal information from micro- and macro-variables in the overall feature network. For instance, the *closeness centrality* expresses the speed at which information goes from one node to another. The *betweenness centrality* represents the control each node exerts over the other nodes in the network. *The eigenvector centrality* expresses the measure of the influence of a node in the network. The *stress* defines a centrality index related to the number of shortest paths passing through a node. The stress score indicates the preferential disposal and distribution paths of dysmorphosis within the system: in biological terms, the stress of a variable indicates the relevance in connecting regulatory nodes (further details on the usefulness of this representation in Chap. 5).

Having a systemic mental model means simplifying the dynamic pattern of reading the succession of events and avoiding getting bogged down in irrelevant details. The systemic thinker judges the behaviour of a system over the long term and not at the level of single events because he is interested in clues about the structure. The structure consists of accumulations, flows, and reinforcing or balancing/unbalancing circuits. It is these that determine the system's behaviour. The structure expresses a process in the visible form, and in the information concerning it, one must see a dynamic configuration.

Information and matter interaction define the material dentofacial system. Gregory and Mary Catherine Bateson define information as "a difference that produces a difference" [63]. Information is fundamental to morphogenesis, from the earliest stages of unicellular and multicellular organisms to facial tissues and organs. Information leads to the structuring of matter (inert matter + Information => structured matter) and the structuring of living systems (structured matter + information => living structures) [64]. Therefore, structure and information provide valuable clues linking past trends and future behaviour. Understanding this biomechanical-functional interdependence helps explain how a narrow defective region can affect others [11, 65–67].

To understand the role of each variable within the patient's developmental history and informational modelling, the orthodontist must also observe the morphological contraction phenomena that attempt to rebalance exorbitant craniofacial areas. The subtraction of supporting capacity may lead to attraction dynamics of neighbouring structures. In the late 1990s, Everet C. Olson (palaeontologist) and Robert L. Miller (geologist) at the University of Chicago focused on the covariation of morphological traits among growing animal structures. They pointed to "*morphological integration*", a central moving force in evolution [67]. The Authors contrasted the importance of the evolution of the *whole* organism against the reductionist view of local anatomical variability, broadly professed by evolutionary biology, palaeontology, and genetics. Anatomical features co-vary spatially over the organism and temporally over the growth periods; parallel to morphology, information and function also evolve. Currently, the degree of integration and covariation among constituents of a biological system (expressed through correlation coefficients, the most meaningful expression of association among morphological traits) tends to increase over time. As the intricate morphological changes that occur during facial growth cannot be independent of each other, studies that do not consider the dependency between evolutionary processes cannot be regarded as exhaustive. However, treatment decisions are also interdependent with each other.

From molecular complexes to tissues to whole organisms, living systems are integrated units with functional and anatomical modular organisation. As shown In Fig. 3.2, Class III malocclusion modules are highly connected traits internally combined by strong interactions but are relatively independent of other modules. Integration between features is more robust and durable over time if concentrated in a single area or dimension, indicating a high degree of correlation between features. Conversely, the correlation between features scattered across all possible sizes will be weak. Studies based on morphological modularity stem from the covariation of traits observed in humans and animals: covariation contains valuable clues as to how morphological characteristics interact during organismal development. The advantage of referring to morphological integration lies in having a conceptual framework that allows the exploration of the relationship between genotype and phenotype.

The variability of living beings during the growth process possesses three basic properties: canalisation, developmental stability, and morphological integration. They describe the *potential* variation, one of many possible system outcomes, not simply the expressed variation. The manifestation of the facial phenotype results from the complexity of disparate factors. It also results from their arrangement within hierarchies and their reciprocal correlations. In the end, and again, it comes down to stating the much-quoted sentence, "The whole is more than just the sum of its parts" [68]. Facial characteristics far outside the average values of a system's components can lead to a theory that may only be valid for a limited period and subjects. Also, the statistical aspects of the theory may vary over time, especially when the context is only partially known, i.e., when there are *latent* characteristics that interact in various ways with the data we have. Most of these interactions introduce errors, distorting statistical procedures.

In the first half of the last century, René Guénon turned his attention to the qualitative aspect of the spatial and temporal determinations of 'organised matter'. In radical criticism of contemporary research that tends to reduce everything to quantitative differentiation and the tendency to redirect everything to uniformity, Guénon speaks of "states of non-manifestation" of living matter. Various levels of reality are juxtaposed in the totality of organisms [69]. René Guénon's thinking contains food for thought for the orthodontist who intends to use computational systems to improve the quality of professional performance. In many situations, algorithms need to compress information and discard non-essential variables to improve the accuracy of the response (more on *Boruta selection* later). This accuracy could be at odds with the lack of complete information about the factors at play in the facial system. Furthermore, it could be helpful to clarify the *counterfactuals* of the processes, i.e., to include in the model the limits of what can happen and the transformations (that could or could not be) realised about the available variables [70].

A doctor striving to define a patient's orthodontic prognosis tries to make sense of events, searching in the complex and unrepeatable organisation of past processes and structures. He questions the possible transformations, the virtuality contained in the phenomena expressed by the facial apparatus. In medicine, there is no single level of explanation of reality, and no story includes all the alternative situations that could have brought about a different outcome. Every pathological state expresses a wide range of evolutionary potential; just as not everything has happened in the past, so not everything will happen in the future. This feature explains the tendency of living beings to behave probabilistically despite the errors and micro-anomalies that develop in organs, which tend to accumulate and upset the forecasts. Each patient represents a specific fusion of biology, randomness, and context. The collected information remains embedded in the system: it is a kind of structural memory that, as we have seen in the skeletal topology of patients where malocclusion worsens with growth, is more stable the more cohesive the structure appears, rich in internal links between the constituents. We will see that morphological information related to a specific type of severe skeletal imbalance is unique to every imbalance. Although changes in evolutionary processes do not co-occur in all local components, the numerical values of angles and segments measured by the cephalometry, in any case, refer to the all-pervasive presence of growth.

As orthodontic processes have a specific structural inaccuracy, in the sense that it is not possible to clarify the totality of the processes involved, this involves a clear, unavoidable statistical error. Some features correspond to relational terms, others to strictly anatomical or geometric terms (although these entities are not separable). Traditionally, we divide the overall morphological structure of the facial apparatus into three main anatomical bases: normofacial, brachifacial, and dolichofacial. The structural pressure of these global configurations strongly influences different local characteristics, such as the spatial arrangement of roots. In the dolichocephalus, the angle between the middle and anterior cranial floors is more open, and the nasomaxillary complex is more protruded.

Moreover, the backward rotation of the mandible and the downward tilted occlusal plane tend to give rise to a tendency for mandibular retrusion. A brachycephalosis-type skull shape is associated with a posterior positioning of the upper jaw, decreased horizontal dimension and relative retrusion of the nasomaxillary complex. The result is a relative forward positioning of the chin.

As shown in Fig. 3.5, when the maxilla tends to grow too much in the sagittal direction (right), the disproportion between the bony bases is absorbed by the mesial movement of the mandibular molars, the palatal retroclination of the upper incisors and the vestibular anteroclination of the lower incisors. If mandibular growth is excessive (left), the upper arch incisors lean buccally, while the lower incisors lean lingually. Additional natural compensatory mechanisms may involve progressive counterclockwise rotation of the occlusal plane and mesial migration of upper tooth elements. The dentoalveolar system surrounding the root elements demonstrates a perceptive ability to assess the discrepancy between the actual situation and the internal map of occlusal equilibrium, like the adaptation of plants to the characteristics of their environment or the animal proprioception. They describe the behaviour of dentofacial components using mathematical formalisation, forcing one to consider

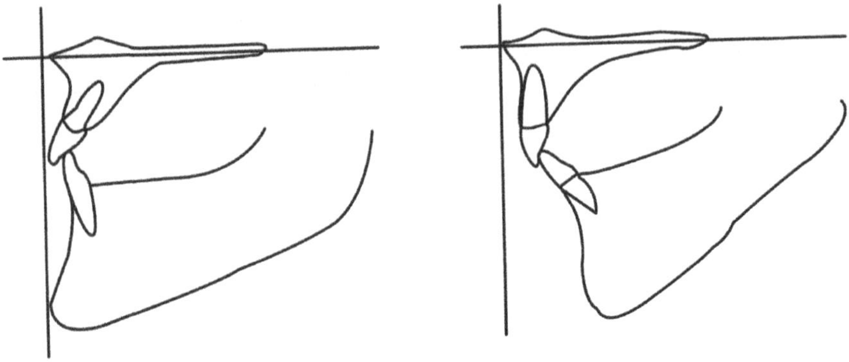

Fig. 3.5 The dento-alveolar compartment has the task of solving complicated problems of congruence between bone bases and dental elements to the benefit of occlusal homeostasis. "Natural" computational systems define these strategies. Honest dento-alveolar compensation in sagittal disproportions in maxillo-mandibular relationships, in overgrowth of the mandible (left) and maxilla (right). In the oral cavity, each tooth element tends to assume the most geometrically convenient position to intercept and dissipate the forces acting on it. Vertical pressures dissipate along the central axis, with minimal engagement of the paradentium

that the essence of the system is relational and adaptive and that the network of relationships takes place within a series of tangible material structures, which specify which structure interacts with which and in which manner. The systems analysis considers the skeletal aspect of each patient together with the dentoalveolar adaptive situation, understood as the degree of general reactivity and the intrinsic inclination towards a concept of occlusal harmony [71]. The facial system has a qualitative idea of space, differentiated in its inclinations and directions. This space can be classified (following Guénon's thought) as 'qualified', capable of showing us its 'primitive simplicity' [9]. Placing each patient within the proper laws of global arrangement and progression means extracting the most appropriate 'prior probabilities' from specific combinations and peculiarities of cephalometric data.

Biological systems create their behaviour based on their structure. Each facial type implies specific "floating norms" of locally acceptable values, growth obstacles, circuits of defect accumulation, and self-reinforcing flows of imbalance or renormalisation that can mask the severity of defects. All these phenomena constitute mental representations that experienced orthodontists perceive as self-evident. Knowledge of the underlying structure of the dentoskeletal data is necessary to identify which system elements are most connected and what kind of influence they exert.

As a starting point, wanting to reduce the facial structure to only two dimensions (following the reconstruction of the facial system performed by cephalometry), it is possible to hypothesise that CoGn-CoA (maxillo-mandibular differential) may represent the most straightforward summary representation of the horizontally-transversely imbalance. In contrast, NMe-CoGo may best represent the imbalance between anterior and posterior facial height, I.e., the global divergence. This coarse-grained two-dimensional pattern makes it possible to appreciate the sum of 'traces' the low-dimensional structure leaves or receives from other organism components. Since the properties of the parts can only be understood based on the organisation of the whole, what predictive clues on growth and treatment outcome can we derive from knowledge of these traces? The CoGn-CoA dimension includes traces from Interincisal, L1MP, U1PP, ANB, Wits appraisal, GoPg-SN, etc. Based on how facial features interact [74], new dimensions are possible to obtain a more comprehensive view of the patient's situation. How can local data be correlated with the three traditional (dolicho-brachi-mesium) skeletal representations? The few-dimensional model allows one to ignore the skein of local iterations and to focus on the aggregate feature effect on the overall skeletal structure [72].

Every orthodontist knows that mitigating a defect does not mean limiting it to one element but rather means rebalancing the relationships that bind disparate parts of the system. The treatment is an attempt to awaken the latent potential of the system, unexpressed due to constraints. The clinician must clarify what was important before and during the process that caused the change. How did it come about? True transformation only occurs in the relations between the parts and their mutual structural pressures. In all living systems, we face contingency's creative and destructive power of unrepeatable singular events. During the growth process, there are periods of relative stability, but the coherences that guide these periods may change over time. No linear and continuous progression of the facial structure can explain how everything that happened was inevitable.

The dentition is a dynamic functional mechanism constantly changing and will continue to adapt throughout the patient's life. The ideal treatment should stimulate and make dominant those links between the hubs with combinatorial abilities opposite to the characteristics expressed in the defect. For example, in the case of an open bite, stimulate ramus growth to encourage mandibular ante-rotation. Or give more support to the vertical dimension of the system to unload the temporomandibular joint. A biological system is an entity that asserts its existence through the relationship between its parts. Balance is the intrinsic goal that the system pursues to protect itself, using continuous feedback mechanisms between the components. Because of the many connections between the different zones, the system tends to resist change and be stable over time. When the orthodontist diagnoses a malocclusion, he arrives "when the damage is done". He can appreciate or intuit the changes the system has gradually made in developing the imbalance. But is he sure that the cause of the defect is where it seems most visible? Although we cannot imagine a single normative trajectory of change, it is possible to identify the most probable end goals of development. The change of each constituent of the facial system can only occur within the possibilities allowed by the state the rest of the system is in. As in a complex system, the cause and effect of the imbalance can be distant in time and space, and both the computational skills and the orthodontist's ability to perceive the patient's actual mapping of processes are necessary.

Systems thinking is a useful conceptual tool and language for describing the latent forces that govern the behaviour of biological systems. The feature interconnections represent the informational part of the systems, which makes them able to amplify and reinforce themselves. Changing the interconnections of a system through a corrective device can lead to a profound change in structure and propensity over time. Some subcomponents might have effects that are decoupled from all others and distort the overall impact of the system or unleash behaviours that were latent in the structure itself. Translating to the growing facial structure, principles of complex competition between dominant-self-reinforcing and balancing circuits allow observation in a dentoskeletal defect progression of an entry, an exit, and a state. Events occur according to dynamic patterns of behaviour. We note specific factors and processes of accumulation and disposal of dysmorphisms, which can be represented schematically (accepting a certain degree of superficiality of the model) in terms of structural pressures and dominance circuits for each type of malocclusion (Box 3.4).

Box 3.4—Dominance Circuits

The current morphological state of the system favours conditions that facilitate the possibilities of different types of circuits, i.e., information flows among features. The assumption is that individual characteristics can best be understood in the form of processes since, in a complex system, the property of each component links to the properties of all the others.: Dominance circuits for Class I malocclusion: information flows in dominances of dental-basal disproportion, maxillomandibular skeletal balance, protrusion of incisive regions, crowding, and 'full mouth' appearance.

Dominance circuits for Class II malocclusion, first division: information flows in dominances of maxillary circuits, excessive overbite, mouth breathing, and tongue posture. Second division: information flows in dominances of deep bite, masticatory muscle hypertone. Dominance circuits for Class III malocclusion: information flows in dominances of mandibular circuits or/and maxillary hypoplasia, densely interconnected structure, intervening processes mitigating/reinforcing the skeletal disharmony, postures, possible teratology, and resistance to nonsurgical therapy.

We are inclined to give greater importance to the forces and quantities entering the facial morphology than the amounts leaving it. This approach is wrong because the sources of resistance to defect correction during treatment often derive from the weak disposal phenomena, which is also the cause of relapse. Systemic thinkers know that in every morphological imbalance, we must also look for the delay in the disposal of that imbalance. And that any feature can be renormalised, or further unbalanced, by nearby structures, just as happens when the simultaneous presence of a high SN-PP angle further aggravates the maxillo-mandibular hyper-divergence caused by a significant increase in the PP-PM angle. Excess of mandibular growth about maxillary growth is often tentatively *dissipated into* the posterior mandibular rotation, which increases the open bite. Anterior growth of the mandible is impeded by the tightening of the deep bite when a Class II malocclusion 2 division is established [73].

Systemic thinking makes it possible to move from the system structure *(stock)* to the system behaviour over time (flux) (see Fig. 3.2) and vice versa. Non-linearity operating in biological complex systems is the leading cause of the passage from one dominance zone to another and the sudden oscillation between an exponential growth related to a reinforcing circuit and a balancing, retroactive circuit, which determines a hypo-dynamic period of the system. These processes often take the form of balancing systems intended to keep the harmony of the various components intact over time. Beyond the statistical importance of specific parameters at stake displayed by the regression procedures such as Relaimpo, we can understand the real significance of the parameter only in light of the constraints and the flows operating in the system. Linear regression procedures predict the target by making a weighted sum of the input characteristics when each feature complies with the condition that the other features remain constant, which is not the case in orthodontics. Somatic information is a vocabulary of changing information. No rule applies for more than a few spaces or periods: rules constantly change, subject to growth, to the randomness of circumstances, and the remodelling and positional displacement of facial constituents. Constraints are far from independent, and growth flows must adapt to them.

We can get an idea of the circuits which characterise complex facial dynamics during the growth process by watching the fluctuation of the skeletal maxillo-mandibular differential CoGn-Coa by observing the interactive global structure of the system (through metrics of Graph analysis) or the network of mutual influences between the components (through Bayesian networks). Most importantly, we must be able to put the answers of the algorithms in the right light, appreciating them from the idea we have made of the systems. We can also run computer simulations, imagining how things might have gone without constraints, even if we don't know the entirety of the reinforcing or balancing constraints present in the "hidden half" of the facial system. Sometimes, the simulation could be inaccurate and misleading because the interconnects are challenging to detect. What *Factor analysis* reveals to us about the latent factors at work in the system is that the horizontal and vertical growth components tend to aggregate to form group-antigroup circuits and dentoalveolar circuits of balancing. (Factor analysis, Figure pag. 173) Problems related to circuits and inapparent parameters are "inherent" in biological systems, influencing their operating units and resilience to change.

Cephalometry is an applied geometry. A geometry that investigates a living, reactant system, revealing important information about the patient's past once placed in the appropriate conceptual framework. The secret lies in knowing how to condition the data to the trend. We can appreciate or imagine the effect of biomechanical translation between contiguous growing skeletal components affecting each other. Still, we cannot appreciate the cumulative impact of some facial components on many others, which determines the morphological homeostasis or the collapse of the whole system. The latter oversees the path from simple rules of local interaction to the emergence of collective, systemic behaviour that governs ventilation, lingual posture, tone of orofacial and post-cervical muscles, arrangement of the bony spicules according to the resultant of the masticatory forces [7]. The orthodontist who reasons in systems analyses the patient's history, whose morphology is the key to understanding *what* is going on and *why* the patient evolved this morphological pattern. At the beginning of the treatment, instead of intervening on the most visible "defective" component, it is better to seek the location of any self-catalytic circuits that bind parts of the system.

By breaking down the system, we lose the overall vision and the meaning of individual characteristics, i.e., the value of that characteristic as if it were a reality. On the other hand, considering a system too broadly can be dispersive. The determinants of features are not only spatial but also temporal. The morphological defect can be related to organogenesis dominant phenomena far away in time (and will probably have consequences in the future). Ignoring this means adopting symptomatic therapeutic solutions: triggers and reinforcing circuits left undisturbed will return. Through network analysis, the systemic orthodontist can identify the connection of the other elements with the one we are analysing and what kind of influence they exert. That is, to know the underlying structure of the data to identify what prevents the system from balancing.

Neglecting to consider these multiple sources at the root of orthodontic problems exposes one to the risk of encountering annoying circumstances, such as:

- I solved a problem, but this very action seems to have created new ones;
- this dysmorphosis has established itself quickly, but its correction is too slow;
- something I thought I had solved strikes back;
- changing the treatment decision in the process made the situation worse.

The dense and subtle network connecting the orthodontic events is invisible to the untrained eye. The system's structure influences the treatment outcome more than our efforts. It is worth "looking around" to solve the problem, to extract the deep dynamics and ways to control them. Systemic thinking is often counterintuitive: it is a circular mental strategy based on a simple thought structure. Computer language has created structures that have made it possible to reduce the gap between the inanimate and the living because they resemble (can represent) many of the things we perceive in natural phenomena. However, this resemblance sometimes needs to be more precise. It all depends on how the process takes place and (once again) how the local components of the process interact.

Constraints operating during organogenesis mean that only a few structures can form, and few morphologies can change after initial formation. Forms relate to the physical medium on which we have our system. As stated, the systems thinker must prioritise interactions over structures to know what is possible in the facial system and what is not. The biological reality realises only one potential among others: the supramolecular coherence constrains the evolutionary bifurcations. The natural laws governing the developmental choices are probabilistic because it is not possible to know what will happen at the evolutionary bifurcation points precisely; the propensity and unfolding programs result when the system behaves as a whole; only then is it possible the different plausibility collates. Perceiving this complex interplay of problems is the task of analytical approaches based on the absence of multicollinearity and endogeneity between variables [29, 74].

When the definitive first molars have just erupted with the head-to-head relationship, a Class I dental matching will be achieved in 75% of cases, while in 22%, a Class II matching will be established, and in 3%, a Class III. Early treatment decisions, once guided by experience and retrospective clinical intelligence, can nowadays be upgraded by computational personalised predictive analytics. But it is realised that even the most up-to-date computational tools, if not correctly set up with reliable, reproducible data with a low degree of randomness, can produce an amplification of prognostic inconsistencies due precisely to the power and the reiteration of the calculation procedures operating on a misunderstood system. The constraint of form acts based on inspective identifiable threshold functions. In a Class III malocclusion, an anterior crossbite correlates with a deep bite condition and self-reinforcing dysmorphia. The temporal stability of the interaction gives rise to a highly robust biological structure in amplifying the imbalance. The opposite occurs in Class II malocclusion 2 division, in which the deep bite and muscle hypertonus constitute an obstacle to the forward progression of the chin.

Systems reasoning makes it possible to appreciate the dominant global behaviour of a structure that persists over time, standard in patients with similar skeletal imbalances subjected to similar evolutionary constraints. In our Class III dataset, we observed that, from age six, all subjects who presented a CoGn/CoA ratio > 1.42 (severe, dominant mandibular circuits) exhibited worsening imbalance as they grew older. In these subjects, parallel to the morphological dysmorphism, some self-reinforcing circuits of imbalance were at work, which imposed dominance on the balancing circuits. A rule emerged that the greater the Class III skeletal imbalance, the more it behaves like an attractor. This rule testifies to the relationship between dominant morphological circuits over the whole propensity of the system.

Another confirmation of the importance of structural constraints over the outcome emerged when splitting a group of patients with this type of malocclusion into three morphological clusters. Among some patients, a vertical imbalance between jaws prevailed (HD, hyper divergence, Fig. 3.6, left) and a tendency to open bite and post-rotation of the mandible. In other patients, a dominant horizontal forward development of the chin about the upper jaw prevailed (HM, hypermandibolia, Fig. 3.6, right). Subjects with HD and HM membership more significant than 60% showed a marked tendency to retain HD and HM characteristics over time. A third morphological cluster of patients was characterised by less severe imbalance characteristics (BAL, balanced features) [75].

The high cluster HD and HM membership constrained these patients to coherent imbalance characteristics over time. In contrast, subjects with mild Class III maxillo-mandibular characteristics (BAL subjects) exhibited a dominance of balancing circuits over self-reinforcing imbalance circuits. This situation exposed the Bal subjects to the risk of being overwhelmed by the sudden predominance of HM or HD circuits during the growth process, often leaving the Bal cluster in favour of HD or HM clusters. Conversely, HD and HM patients with high membership did not leave their cluster. Still, they often increased the degree of their membership, i.e. they acquired additional characteristics or reinforced those they already possessed. In the following chapters, we will see the capability of fuzzy cluster analysis to

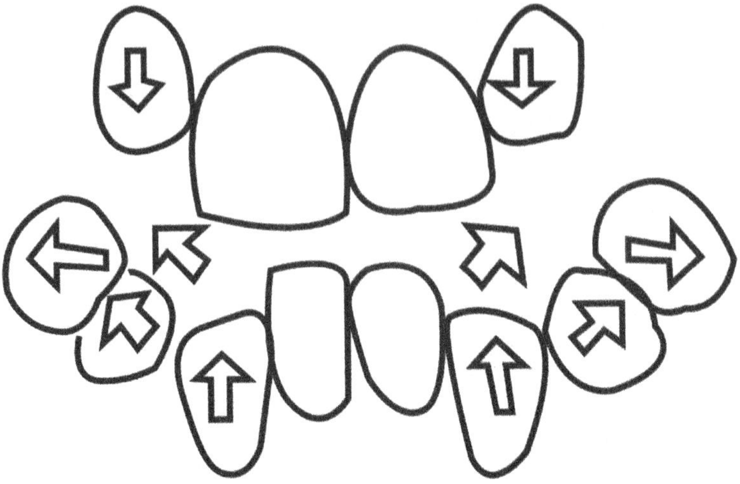

Fig. 3.6 Eruptive paths of incisive teeth and compensation mechanisms to favour the proper structure. In the mandible, diastemas are closed by the eruptive force of the lower first permanent molars. The germ of the upper first permanent molar develops in the maxillary tuberosity with an eruptive course downward and backward, promoting normal Class I occlusion. Small spatial and temporal changes at critical periods of the eruptive process can lead to significant differences in final occlusal conditions. Small spatial and temporal changes at essential periods of the eruption process can lead to substantial differences in the absolute conditions. Local phenomena are strongly influenced by the global characteristics of the system, i.e. by the mutual relationship of the bone bases. Readapted from: [86]

define the strength of the HM and HD imbalance supply circuits according to the percentage of membership to the cluster. Considering the information content that can enter a predictive system, a patient with a high membership in one's cluster must be regarded as a 'qualified' patient, i.e., particularly expressive of a stable clinical condition we can use for machine learning. Defining membership decreases the extent of developmental noise, the mechanisms amplifying random growth perturbations.

In physics, to state that the behaviour of a system has changed based on the information received is equivalent to saying that forces have acted on the system. As forces do, uncertainty also propagates through skeletal circuits, the knowledge of which is decomposable and visible in a network of covariates in which other variables influence each variable. When variables are not independent, the correlation between two attributes brings out new attributes, making it possible to reason for *stable* aggregate data (from cross-sectional patients) or *dynamic* aggregate data (from patients followed over time). Forecasting means identifying favourable and unfavourable factors when they have yet to materialise. The possibility of having a spatial filter of correlations between variables generates a 'functional map' derived mathematically from the spatial proximity of features and anatomical or functional characteristics. This map made it possible to establish that the predictive accuracy is superior in some parts of the data space than in others [76]. The path leads from averages to the personalisation of prognostic reasoning.

Each component of the facial system likely acts according to a threshold model: each element has a threshold and can activate whenever the stimulus exceeds it. Nodes with lower activation thresholds respond first to stimuli ('early adopters'), conditioning neighbouring nodes through positive feedback loops. The emergent property of the system is an effect of the propagation of a local disharmony counterbalanced by the modification of the myriad interactions between the system's components. Therefore, the system becomes the true counterpart of provincial disharmony. Once the dentofacial system reaches the critical condition of occlusal instability, It can *percolate* into collective dynamics capable of changing the scenario relatively quickly. In these circumstances, the coarse graining of a system could be desirable, even though the process of abstraction could eliminate essential details expressive of system dynamics.

When a predictive algorithm learns, all it does is iteratively find the best parameters to make the model adhere to the reality of the input orthodontic data, albeit ever-changing. For guiding the learning of rules, the orthodontist receives information from the data, uses a program, and ensures that the machine can receive the maximum amount of information about the patient's clinical background. The learning data set may consist of a collection of cases or a single patient.

As stated, the core of understanding in systems orthodontics depends on searching for *global organising principles* rather than models that outline the evolution of single parameters. In addition to a careful clinical evaluation of the dentition, orthodontic definition and treatment design require assessing the patient's functional environment and its most likely evolution over time. The orthodontist knows that the assessment of the problem presented by the patient stems from morphological characteristics that are bound to change due to growth and the effect of corrective devices. The boundaries of the future variability are challenging to assess a priori: what we see must be judged by what happened in the period before and what is likely to occur in the future. Each patient represents a piece of information whose values represent the specificity of the individual patient. Applying uniform diagnostic procedures to unique situations fatally leads to misunderstandings since, in orthodontics, the exact morphogenetic cause can produce unpredictably different effects.

Stability refers to the relative position of the patient features about the average in the flow of time. Some observations may depend on each other due to some underlying process. The 'antecedent-connective-consequent' feature covariation paradigm may become invalid by changing a consequent variable affecting a concurrent antecedent variable. Multiple independent features can use numerous independent sources of information about the outcome. The answers that computational machines offer us after layers of reiteration are increasingly short-lived abstractions. Moreover, beyond the formal clinical-mathematical logic, information about the physical substrate on which facial growth operates is also critical and must be preserved during statistical operations. The abstraction procedure detects specific qualitative patterns and extracts high-level concepts from time-stamped data.

Learning means deducing the most from each observation by tracing the observations back to their causes [77]. In formulating a forecast, the human brain possesses an innate procedure of combinations of symbolic thoughts, potential hypotheses, and intuitions about the hypothetical future. In everyday life, we pick up signals from the world and select models of possible and plausible scenarios. We can also update them based on data from experience. We can change opinions and, in time, make more mature judgements about many of life's events. We have an innate ability to hypothesise causal links and to find causal clues in simple correlations between phenomena, disregarding the problematic Nature of the processes and our limited knowledge of all nuances of phenomena.

Our uncertainty decreases as we make new clinical observations: the probability we attribute to the world's facts is the expression of our residual uncertainty. As mentioned, most orthodontists' operational decisions are made based on incomplete and approximate rule-based premises, for more, affected by a fair number of exceptions. The orthodontist is inclined to formulate causal connections, trusting that they have a sufficiently thorough knowledge of the present phenomena to predict future development scientifically.

Computational resources help orthodontists in the hope of extracting knowledge from increasingly complex patient data sets. When an algorithm learns, all it does is iteratively find the best expressive combination of the parameters to make the model adhere as best as possible to the non-numerical reality of the input data. Each concept relates to the patient's dominant properties (habits, whole morphology, growth potential, extent of circuits of imbalance and renormalisation, and so on). Faced with the combinatorial effect of the components that generated the specificity of craniofacial morphology, the orthodontist must behave like a naturalist discovering each time a new continent: he finds that living matter possesses its capacity to perform specific forms of *morphological calculations* to achieve its purposes and to act on the results of these calculations. The promise is to unveil the true essence of phenomena beyond their appearance: the facial apparatus exhibits an intrinsic ability to process and integrate different information from the outside (positional, relational, environmental), using it to regulate its development inside patterns. The system tends to keep itself in harmony with its expectations of the environment; in the same way, the dental arches can process information, extract and sift data from the surrounding environment, and prepare the dentoalveolar compartment to optimise the dental arrangement biomechanically. The system must be a computational device to achieve all of this.

Computation is, in a broad sense, how a biological system uses information to accomplish a task, adapt to the environment, or improve its fitness. Information flows into the system through the collective action of many components. Information is delocalised: no single component has all the necessary information to influence the whole system's behaviour profoundly. However, we have seen that some facial features tend to come together in clusters because of their strong interaction to form "attractors" and feature subsets ("motifs" or "modules") [78] with a substantial impact on the overall system. We must focus on these information-rich areas to grasp the system's propensity.

In biology, things evolve following laws of growth. As always happens in complex systems, the rules governing facial development are elusive and difficult to generalise, capriciously sensitive to the specificities of each patient. Every organisational change influences a morphological change and, at the same time, derives from it. It must be systemic in Nature to be classified as real change [41]. Any local change must be consistent with the overall system to maintain unity. Thus, change must occur "within the set of constraints and possibilities allowed by the states in which the rest of the system finds itself" [41].

Craniofacial growth involves a delicate interaction between the transformations of the skeletal proportions, muscles, masticatory, swallowing, speaking and mimic functions. This process is exposed over time to favourable/unfavourable influences, habits, lifestyles, and evolutionary phenomena in terms of distortions of the bone bases, crowns and root dystopias, crowding, and other local and global phenomena. The orthodontist is part of this process by superimposing external biomechanical force on endogenous processes using light wires (which act as energy absorption modules of the system), rigid wires (which operate as energy distribution entities), and maintenance devices. It is a continuously reiterated diagnosis and therapy exercise applied throughout the relationship with the patient. Usually, orthodontists divide the list of problems into three components (skeletal pattern, face, dental arches), aiming to linearise and standardise complex dynamics into basic diagnostic and therapeutic concepts and mitigate the complexity of the corrective process. The orthodontist must provide the machine with the possibility of discriminative learning that can specify the difference between development (understood as a progressive form of change) and treatment effect as local connections between stimulus and response. Data must be different and represent the multidimensional and pervasive Nature of development to achieve this learning. We must focus on the specificity of individual processes and the effects of treatment through a discriminatory comparison between processes and morphologies that includes a judgement of the quality of events and phenomena.

Some forms of intra-individual feature variability do not constitute a fundamental change, for example, the body temperature variability throughout the day. The overall structure remains stable; there is no perceptible change in interdependence among parts of the organism. As a first approximation, it is possible that, given the specific characteristics of the previous states and the system's current state, a particular set of changes is likely to occur. Albeit with a probabilistic character, so with a certain degree of error in the prediction. Computational tools cannot accurately predict phenomena unfolding, but they can narrow the uncertainty, narrowing the margins of error.

We must deal with macroscopic dynamical processes to serve as theoretical frameworks for orthodontics. Unlike physical data, orthodontic data (as all biological data) have a reciprocal spatiotemporal combinatorial complexity. Thus, each feature belongs to an underlying anatomical/functional structure, and it has a weight describing the importance of the function to which it is linked. Each cephalometric feature has a specific field of influence on other features and a kind of two-dimensionality: numerical values (mm, degrees) and the influences it causes or receives from neighbouring and distant nodes.

One of the most penetrating insights of computational methods lies in their ability to visualise the relationship between structural information and dynamic behaviour. Computational machines help us to see the growing facial system in terms of conserved stable evolutionary processes, emergent spreading of variety, nucleations of disharmonies, and adaptation phenomena. Moreover, they can provide the confidence level for each observation. What patient features are is defined by the context, which also includes the individual history, either developmental or physiopathological: it is only in the personalised retrospect that we can recognise the significant founding events of disharmonies [79]. The organisational network of these connections can reveal relationships of a purely geometric nature (the relationship that binds the corners of Tweed's triangle), unexpected functional links between specific *dominant nodes* in the system and locating groupings of nodes endowed with more significant available activity.

As stated, the global coherence of reciprocal connections defines the structure. Given that growth is a phenomenon of instability, dysmorphic phenomena can behave like local "crystallised accidents", accumulating information affecting the entire course of events. The parts lose their individuality in the system; on the other hand, the system becomes unity despite the diversity of parts. To consider the whole in terms of its parts and the reciprocal is to recognise that the nonlinear interaction of the parts favours the emergence of new qualitative properties. We can investigate these unique properties, provided their integration with the overall organisational principles of the system is defined. In different biological systems, the actual role of the parts emerges when considering the influence exerted upon them by the whole. Thus, the whole and the parts reciprocally determine each other's function.

Denis Noble referred to these phenomena as the 'Theory of Biological Relativity': things only make sense about context [44]. As an example, the pacemaker rhythm of the heart is not only related to the activity of ion channels. Still, it also depends on the organ's functioning and the whole body. "…Higher levels in biological systems exert their influence over the lower levels. Each level provides the boundary conditions under which the processes at lower levels operate. Without boundary conditions, biological functions would not exist" [44]. This approach puts the organism into the picture and emphasises the importance of living matter's capacity for integration and regulation between multilevel components [44, 80]. The two fundamental organisational relationships (the parts-whole and the inside-outside relations) are the foundation that makes each living organism an autonomous system capable of its survival and adequate response to environmental changes [81]. For the organism to maintain its harmony over time, the parts *must* be by the operational norms of the whole. The behaviour of the parts will be judged 'satisfying' or 'unsatisfying' depending on whether they contribute to collaboration or obstruction in meeting the system's needs.

Since the parts are nothing more than the temporary manifestation of the organisational unity of the whole, *the whole exhibits a greater degree of invariance over time than its parts* "…because it maintains its organisation by continuously breaking down and replacing the constituents that make it up. Consequently, it makes no sense to identify an organisation with the sum of its parts, as these are constantly being replenished by the whole" [80]. In the case of an anterior open bite, the occlusal energy is weakly transmitted through the cusp slopes from one arch to the other. Even in excessive overjet, the compensation mechanisms are not valid due to poor interocclusion of the arches. Good occlusion restores compaction needed to re-harmonise the bone bases. In the case of Class II malocclusion, the compensation patterns appear as occlusal plane rotated clockwise and vestibulo-version of lower incisors. Incidentally, malocclusions of Class III with hyperdivergence, as well as Class II with hypodivergence, are all situations that make the compensation process difficult.

Systems reasoning pushes us to consider causality between events similar to reciprocal causality. If A causes B, it is also possible that B is the cause of A. Any change within A or B corresponds to a change in the connected variables. In patients suffering from Class III malocclusion, the Delaire mask exerts a therapeutic effect related to the "drawer" movement of the upper jaw and the modification of the mandibular condyle. However, it receives a biomechanical counter-effect from the patient's dentoskeletal structure. The latter returns to the mask a counter-effect proportional to the extent of the horizontal or vertical maxillomandibular imbalance. The therapeutic effect can be very different among patients, even when applying the same elastic force,

Systems thinking seeks to represent structures of relationships at work in the causal fields of biological systems [62]. It shares many of its conceptual roots with those of a recent model of thinking underlying the Biomimicry approach, which aims to delve into the processes set in motion by Nature to find solutions [82]. We must regard natural organisms and systems as "masters, creative by necessity", who already solved billions of years ago the problems we strive to solve today. Also, in the logic of Biomimicry, the guiding idea is that in biological systems in perpetual change, organisational principles matter more than individual constituents.

The conceptual shift from reductionism to a holistic orthodontic perspective minimises the loss of predictive accuracy in growing biological systems caused by the slow fading of initial characteristics over time, which characterised the treatment plan. Appropriately fed with data expressive of the problem at hand and its evolution, computational algorithms adapt to orthodontic feature drift by transforming and projecting concepts into a coherent background of the patient's dentoskeletal progressive possible processes of reframing, distortion, and normalisation. New bio-mathematical concepts are minimising what had seemed to be the non-scientific residues of human sciences: uncertainty, non-reproducibility, plurality, and contradiction. Starting from the assumption that variability is the signature of complexity in the living world, we should not seek normality in the single ideal model but in the levels of adaptation, in the balance of compensations between phenomena, and the inclusive architecture of natural ergonomics [83]. Just as complex living systems do, computational architectures applied to deciphering biological phenomena can learn and refine their action through continuous training. Those processes must mimic the action of Nature. For our purposes, we may say that Nature:

- Abhors waste; it uses only the energy it needs.
- Has cyclical processes.
- Rewards cooperation
- invests in diversity.
- Requires local knowledge to which it resorts to resolve external aggression.

- Curbs excesses from within
- harness the power of limits.

(Modified, from [82])

In their daily professional reality, the orthodontist must deal with various sources of data and multiple sources of noise. Each contains different types of information: they must perform interactive data management at various levels of abstraction, ensuring that the different levels of abstraction fit together meaningfully. No knowledge or information stands alone, so the orthodontist must provide a cognitive background that can interpret and effectively understand the various levels of knowledge emerging from data. The point of the matter is that there still needs to be a more in-depth understanding of the processes that generate the different types of variation and malocclusion onsets. However, some empirical-computational laws help narrow the uncertainty of what may happen in the future for patients. Long-term overall growth phenomena can be influenced and sometimes obscured by short-term facial growth phenomena. Long-term phenomena usually affect the global structure of the system, while short-term processes affect local phenomena. Structures that persist over time get their connection by robust co-evolutionary relationships, visualised at least in part in the topology of this system.

Early growth acceleration has gained importance since the multiple factors influencing facial growth in preadolescence began to be elucidated [84]. The facial system evolves following a chain of events: some skeletal zones grow before others, dragging other zones behind them or forcing neighbouring zones to change their characteristics. When the mandibular ramus grows vertically more than the alveolar compartment below the molars, the mandible is moved into an ante-rotation to preserve a tight occlusal relationship. This phenomenon causes the bite to close. On the contrary, if the alveolar bone below the molars grows more than the mandibular ramus, the jaw can only post-rotate, and the bite opens. There is no way to make reliable predictions about these dynamics based solely on the starting anatomical situation of the patient related to local features. However, knowledge of the overall propensity of the system and the degree to which it belongs to a divergence or horizontal cluster can narrow the margins of uncertainty. We must imagine growth as a sequence of information, where the results of each step become the inputs of the next step. We must have a system map; otherwise, since the succession of steps is not uniform in all patients, different developmental bifurcations may lead to different outcomes [85].

Suppose a patient has data concerning a high membership of some form of imbalance. In that case, one can be confident that the patient will present a linear growth trajectory, i.e., with little change in their initial characteristics. For example, a subject with a high cluster membership to skeletal maxillomandibular hyperdivergence (HD) will not develop mandibular ante-rotation. Membership information allows reasonably robust assumptions about the probable features of future facial behaviour based on what has happened in the past. Unfortunately, this rule is valid for Class III patients with a well-defined horizontal (HM) or vertical (HD) type of skeletal imbalance. There is currently insufficient data to confirm or deny the rule observed among Class III patients in other malocclusions.

Table 3.1 After two years, patient A showed a marked worsening of the imbalance, with Wits appraisal values rising from −4.5 to −9.4, despite the dentoalveolar compensation mechanisms put in place (L1MP from 73.9 to 64.6, U1PP from 91.6 to 110.2 degrees). In contrast, patient B slightly improved during two years of growth (this occurs in no more than 25% of patients with Class III malocclusion). The Wits appraisal went from −3.6 to −2.2, and the overjet from −2.1 to 0.6

	ANB	CoA	CoGn	L1MP	OJ	PPMP	PPSN	SNA	SNB	U1PP	SNGoGn	Wits
A	−1.2	68.3	92.4	73.9	−0.1	29.8	14.0	68.9	70.1	91.6	43.8	−4.5
B	2.5	70.6	90.5	92.0	1.7	24.1	10.3	80.6	78,0	110,2	33.8	−2.2

Cephalometric analysis is a valuable descriptive technique that indicates the patient's overall dentocephalic configuration and suggests where significant structural imbalances appear (these zones can function as growth attractors). The thoughtful observation of the child's facial bio-architecture sometimes leads one to suspect the establishment of early signs of malocclusion that are bound to worsen. Severe size disproportion between teeth and bone bases anticipates an early defective fit between dental arches due to the impossibility of achieving a harmonious match in the maxillofacial complex. Frequently, simultaneous to this situation is a disturbance of the eruption sequence (Fig. 3.6). Dentoalveolar compensation mechanisms can reduce the effect of these disharmonies. Some orthodontists believe that deficits in compensation mechanisms rather than structural discrepancies between the maxilla and mandible cause malocclusion.

References

1. K. Kelly, in *Out of Control: the new biology of machines, social systems, and the economic world*. Addison-Wesley, 1995. Accessed: Dec. 12, 2023. [Online]. Available: https://books.google.it/books/about/Out_of_Control.html?id=mGZfZKtsxtwC&redir_esc=y
2. A.L. Barabási, N. Gulbahce, J. Loscalzo, Network medicine: a network-based approach to human disease. Nat. Rev. Genet. **12**(1), 56–68 (2011). https://doi.org/10.1038/nrg2918
3. T. Ritchey, in *Wicked Problems—Social Messes* (Springer Berlin Heidelberg, Berlin, Heidelberg, 2011). https://doi.org/10.1007/978-3-642-19653-9
4. J. Kempf, N. Siegmund, in *Visualizing General Morphological Analysis via Multidimensional Scaling Chair of Software Product Lines Chair of Software Product Lines* (University of Passau, 2015). Accessed: Oct. 07, 2023. [Online]. Available: https://www.se.cs.uni-saarland.de/theses/JuliusKempfBA.pdf
5. R.S. Lee, F.J. Daniel, M. Swartz, S. Baumrind, E.L. Korn, Assessment of a method for the prediction of mandibular rotation. Am. J. Orthod. Dentofacial. Orthop. **91**(5), 395–402 (1987). https://doi.org/10.1016/0889-5406(87)90392-1
6. J.E. Cohen, Mathematics is biology's next microscope, only better; biology is mathematics' next physics, only better. PLoS Biol. **2**(12), e439 (2004). https://doi.org/10.1371/journal.pbio.0020439
7. S.E. Bishara, in *Textbook of Orthodontics* (Saunders, 2006). Accessed: Oct. 07, 2023 [Online]. Available: https://www.amazon.it/Textbook-of-Orthodontics/dp/8181470729
8. R. Freksa, C. Röhrig, Dimensions of qualitative spatial reasoning, in *Qualitative Reasoning in Decision Technologies*, ed. by M.G.S.N. Piera Carreté (CIMNE Barcelona, 1993)
9. R. Guénon, Le règne de la quantité et les signes des temps (Gallimard, 1945)

10. L. White, Early orthodontic intervention. Am. J. Orthod. Dentofacial. Orthop. **113**(1), 24–28 (1998). https://doi.org/10.1016/S0889-5406(98)70273-2
11. D. Sornette, Predictability of catastrophic events: material rupture, earthquakes, turbulence, financial crashes, and human birth. Proceed. Nat. Acad. Sci. **99**(suppl_1), 2522–2529 (2002). https://doi.org/10.1073/pnas.022581999
12. J. Mah, Predictive orthodontics: a new paradigm in computer-assisted treatment planning and therapy. Semin. Orthod. **8**(1), 2–5 (2002). https://doi.org/10.1053/SODO.2002.28164
13. L. Perillo et al., Machine learning in the prognostic appraisal of Class III growth. Semin. Orthod. **27**(2), 96–108 (2021). https://doi.org/10.1053/J.SODO.2021.05.006
14. A.R. Golding, P.S. Rosenbloom, Improving accuracy by combining rule-based and case-based reasoning. Artif. Intell. **87**(1–2), 215–254 (1996). https://doi.org/10.1016/0004-3702(95)00120-4
15. P. Auconi, E. Ottaviani, E. Barelli, V. Giuntini, J.A. McNamara, L. Franchi, Prognostic approach to Class III malocclusion through case-based reasoning. Orthod. Craniofac. Res. **24**(S2), 163–171 (2021). https://doi.org/10.1111/ocr.12466
16. A. Aamodt, E. Plaza, Case-based reasoning foundational issues, methodological variations, and system approaches. AI Comm. **7**, 39–59 (1994). Accessed: Jan. 03, 2023 [Online]. Available: https://scirp.org/reference/referencespapers.aspx?referenceid=2816811
17. E. Hüllermeier, in *Case-Based Approximate Reasoning (Theory and Decision Library B, 44)*, 2007th edn. (Springer, 2013)
18. A. Holt, I.B. Ritz, R. Schmidt, P. Perner, Medical applications in case-based reasoning. Knowl. Eng. Rev. **20**(3), 289–292 (2005). https://doi.org/10.1017/S0269888906000622
19. M. Blastland, in *The Hidden Half: The Unseen Forces that Influence Everything* (Atlantic Books, 2020)
20. N.R. Sanders, L.P. Ritzman, in *Judgmental Adjustment of Statistical Forecasts* (Springer, Boston, MA, 2001), pp. 405–416. https://doi.org/10.1007/978-0-306-47630-3_18
21. S. Finlay, *Predictive Analytics, Data Mining, Big Data* (Palgrave, 2014)
22. N.V. Chawla, K.W. Bowyer, L.O. Hall, W.P. Kegelmeyer, SMOTE: synthetic minority over-sampling technique. J. Artif. Intell. Res. **16**, 321–357 (2002). https://doi.org/10.1613/jair.953
23. E. Siegel, in *Predictive Analytics: The Power to Predict Who Will Click, Buy, Lie, or Die* (Wiley & Sons, New York, 2016). Accessed: Dec. 12, 2023 [Online]. Available: https://www.wiley.com/en-ie/Predictive+Analytics%3A+The+Power+to+Predict+Who+Will+Click%2C+Buy%2C+Lie%2C+or+Die%2C+Revised+and+Updated-p-9781119145677
24. M.A. Bedau, Can unrealistic computer models illuminate theoretical biology?, in *1999 Genetic and Evolutionary Computation Conference Workshop Program*, ed. A. Wu (1999), pp. 20–23
25. S.A. Levin, B. Grenfell, A. Hastings, A.S. Perelson, Mathematical and computational challenges in population biology and ecosystems science. Science **275**(5298), 334–343 (1997). https://doi.org/10.1126/science.275.5298.334
26. S.N. Asiri et al., Applications of artificial intelligence and machine learning in orthodontics. APOS Trends Orthodont. **10**(1), 17 (2020). https://doi.org/10.25259/APOS_117_2019
27. Y.M. Bichu, I. Hansa, A.Y. Bichu, P. Premjani, C. Flores-Mir, N.R. Vaid, Applications of artificial intelligence and machine learning in orthodontics: a scoping review. Prog. Orthod. **22**(1), 18 (2021). https://doi.org/10.1186/s40510-021-00361-9
28. P. Auconi, et al., The validity of machine learning procedures in orthodontics: what is still missing? J. Pers. Med. **12**(60 (2022). https://doi.org/10.3390/jpm12060957
29. V. Allareddy, S.R. Venugopalan, R.P. Nalliah, J.L. Caplin, M.K. Lee, V. Allareddy, Orthodontics in the era of big data analytics. Orthod. Craniofac. Res. **22**(1), 8–13 (2019). https://doi.org/10.1111/ocr.12279
30. H. Haken, in *Erfolgsgeheimnisse der Natur Synergetik: Die Lehre vom Zusammenwirken* (Deutsche Verlags-Anstalt, 1994). Accessed: Oct. 07, 2023. [Online]. Available: https://www.buchfreund.de/de/d/e/9783499197444/erfolgsgeheimnisse-der-natur-synergetik-die-lehre?bookId=102765412
31. F. Jullien, *Conférence sur l'efficacité* (1951). Accessed: Oct. 07, 2023. [Online]. Available: https://www.puf.com/content/Conférence_sur_lefficacité_0

32. G. Bateson, *Mind and nature a necessary unity* (1979), Accessed: Oct. 07, 2023. [Online]. Available: http://www.oikos.org/mind&nature.htm
33. M. Brodsky, Partial redundancy and morphological homeostasis: reliable development through overlapping mechanisms. Artif. Life **22**(4), 518–536 (2016). https://doi.org/10.1162/ARTL_a_00216
34. A. Blum, T. Mitchell, in *Combining Labeled and Unlabeled Data with Co-training*. Proceedings of the eleventh annual conference on Computational learning theory, New York, NY, USA. ACM (1998), pp. 92–100. https://doi.org/10.1145/279943.279962
35. J.S. Casko, W.B. Shepherd, Dental and skeletal variation within the range of normal. Angle Orthod. **54**(1), 5–17 (1984). https://doi.org/10.1043/0003-3219(1984)054<0005:DASVWT>2.0.CO;2
36. R.M.S. Hammond, T.J. Freer, Application of a case-based expert system to orthodontic diagnosis and treatment planning. Aust. Orthod. J. **14**(4), 229–234 (1997). https://doi.org/10.3316/INFORMIT.976744483304914
37. E. Freer, T.J. Freer, Variations in treatment need using four screening methods. Aust. Orthod. J. **15**(4), 214–218 (1999). Accessed: Oct. 07, 2023. [Online]. Available: https://doi.org/10.3316/INFORMIT.986750388870567
38. K. Zimmermann, C. Freksa, Qualitative spatial reasoning using orientation, distance, and path knowledge. Appl. Intell. **6**(1), 49–58 (1996). https://doi.org/10.1007/BF00117601
39. T.A. Russ, Using hindsight in medical decision making. Comput. Methods Programs Biomed. **32**(1), 81–90 (1990). https://doi.org/10.1016/0169-2607(90)90088-q
40. N. Lavrač, E. Keravnou, B. Zupan, Intelligent data analysis in medicine. J. Comput. Sci. Technol. (2000). Accessed: Dec. 13, 2023. [Online]. Available: https://www.semanticscholar.org/paper/Intelligent-Data-Analysis-in-Medicine-Lavra%C4%8D-Keravnou/3e4f8a3056aa7d8e5947da6f071e2515aa9225f1
41. D.H. Ford, R.M. Lerner, Developmental systems theory: an integrative approach, 1st edn (SAGE, Newbury Park, CA, 1992)
42. F. McDonald, A.J. Ireland, in *Diagnosis of the Orthodontic Patient* (Oxford University Press, 1998). Accessed: Oct. 07, 2023. [Online]. Available: https://global.oup.com/academic/product/diagnosis-of-the-orthodontic-patient-9780192628893?cc=it&lang=en
43. P. Fudalej, M. Dragan, B. Wedrychowska-Szulc, Prediction of the outcome of orthodontic treatment of Class III malocclusions–a systematic review. Eur. J. Orthod. **33**(2), 190–197 (2011). https://doi.org/10.1093/ejo/cjq052
44. D. Noble, A theory of biological relativity: no privileged level of causation. Interf. Focus **2**(1), 55–64 (2012). https://doi.org/10.1098/rsfs.2011.0067
45. G.F. Chew, 'Bootstrap': a scientific idea? Science **161**(3843), 762–765 (1968). https://doi.org/10.1126/science.161.3843.762
46. F. Capra, in *The Tao of Physics: An Exploration of the Parallels Between Modern Physics and Eastern Mysticism* (Shambhala Publications, 1991). Accessed: Feb. 12, 2024. [Online]. Available: https://www.google.it/books/edition/The_Tao_of_Physics/NmhfQgAACAAJ?hl=it
47. Y. Hu, N. Kasabov, W. Liang, Personalized information modeling for personalized medicine. In: Springer Handbook of Bio-/Neuroinformatics (Springer Berlin Heidelberg, Berlin, Heidelberg, 2014), pp. 533–553. https://doi.org/10.1007/978-3-642-30574-0_33
48. N. Kasabov, in *Evolving Connectionist Systems: The Knowledge Engineering Approach* (Springer, London, 2007). https://doi.org/10.1007/978-1-84628-347-5
49. R. Benkirane, in *La complexité, vertiges et promesses : 18 histoires de sciences* (Le Pommier, 2002). Accessed: Oct. 07, 2023. [Online]. Available: https://www.amazon.it/complexit%C3%A9-vertiges-promesses-histoires-sciences/dp/2746501058
50. T. Dethlefsen, in *Schicksal als Chance: das Urwissen zur Vollkommenheit des Menschen* (Goldmann, 1980). Accessed: Oct. 07, 2023. [Online]. Available: https://www.amazon.de/Schicksal-Chance-Urwissen-Vollkommenheit-Menschen/dp/3442117232
51. L. von Bertalanffy, in *General System Theory: Foundations, Development, Applications* (G. Braziller, 1968). Accessed: Dec. 13, 2023. [Online]. Available: https://repository.vnu.edu.vn/handle/VNU_123/90608

52. G. Lando, in *Mereology: A Philosophical Introduction* (2018)
53. A. Bejan, in *Advanced Engineering Thermodynamics* (Wiley, 1997). Accessed: Oct. 07, 2023. [Online]. Available: https://www.amazon.it/Advanced-Engineering-Thermodynamics-Adrian-Bejan/dp/0471148806
54. W. Rudin, in *Real and Complex Analysis*, 3rd edn. (McGraw Hill, 1987). Accessed: Oct. 07, 2023. [Online]. Available: https://59clc.files.wordpress.com/2011/01/real-and-complex-analysis.pdf
55. P. Vineis, in *Nel crepuscolo della probabilità: la medicina tra scienza ed etica* (Einaudi, 1999). Accessed: Oct. 07, 2023. [Online]. Available: https://www.einaudi.it/catalogo-libri/scienze/medicina/nel-crepuscolo-delle-probabilita-paolo-vineis-9788806147181/
56. G. Bateson, in *Men are Grass: Metaphor and the World of Mental Process—Gregory Bateson—Google Libri* (Lindisfarne Association, 1980). Accessed: Dec. 13, 2023. [Online]. Available: https://books.google.it/books/about/Men_are_Grass.html?id=gWaOHAAACAAJ&redir_esc=y
57. Y. LeCun, Y. Bengio, Convolutional networks for images, speech, and time series, in *The Handbook of Brain Theory and Neural Networks*, vol. 3361 (1995), pp. 255–258. Accessed: Dec. 13, 2023. [Online]. Available: https://philpapers.org/rec/LECCNF
58. D. Gentner, V. Gunn, Structural alignment facilitates the noticing of differences. Mem. Cognit. **29**(4), 565–577 (2001). https://doi.org/10.3758/bf03200458
59. D.H. Enlow, T. Kuroda, A.B. Lewis, Intrinsic craniofacial compensations. Angle Orthod. **41**(4), 271–285 (1971). https://doi.org/10.1043/0003-3219(1971)041<0271:ICC>2.0.CO;2
60. T. Siegfried, in *A Beautiful Math* (Joseph Henry Press, Washington, DC, 2006). https://doi.org/10.17226/11631
61. D.G. Hernández, D.H. Zanette, Critical transition induced by neighbourhood size in evolutionary spatial games. Eur. Phys. J. B. **82**(3–4), 361–366 (2011). https://doi.org/10.1140/epjb/e2011-20277-8
62. D. Meadows, in *Thinking in Systems. A Primer* (Sustainability Institute, Chelsea Green Pub. Co., 2008). Accessed: Oct. 07, 2023. [Online]. Available: https://www.iseesystems.com/store/books/thinking-in-systems/preview.pdf
63. G. Bateson, M.C. Bateson, Angels Fear: towards an epistemology of the sacred (2005). Accessed: Feb. 17, 2024. [Online]. Available: https://books.google.it/books/about/Angels_Fear.html?id=hzsvPwAACAAJ&redir_esc=y
64. F. Gaiseanu, Informational structure of the living systems: from philosophy to informational modelling. Philosop. Study **10**(12) (2020). https://doi.org/10.17265/2159-5313/2020.12.004
65. M. Strevens, Bigger than chaos: understanding complexity through probability. Choice Rev. Online **41**(5), 41–2867 (2004). https://doi.org/10.5860/choice.41-2867
66. J.T. Richtsmeier, V.B. Deleon, Morphological integration of the skull in craniofacial anomalies. Orthod. Craniofac. Res. **12**(3), 149–158 (2009). https://doi.org/10.1111/j.1601-6343.2009.01448.x
67. E.C. Olson, R.L. Miller, in *Morphological Integration* (University of Chicago Press, 1999). Accessed: Oct. 07, 2023. [Online]. Available: https://press.uchicago.edu/ucp/books/book/chicago/M/bo3620375.html
68. K.E. Willmore, N.M. Young, J.T. Richtsmeier, Phenotypic variability: its components, measurement and underlying developmental processes. Evol. Biol. **34**(3–4), 99–120 (2007). https://doi.org/10.1007/s11692-007-9008-1
69. R. Guénon, *Les États multiples de l'être* (Édition Vega, Paris, 1932)
70. C. Marletto, in *The Science of Can and Can't: A Physicist's Journey Through the Land of Counterfactuals*. (Penguin Books, 2021). Accessed: Oct. 06, 2023. [Online]. Available: https://www.hoepli.it/libro/the-science-of-can-and-can-t/9780141986463.html
71. I.L. Nielsen, Analysis of general facial growth, maxillary and mandibular growth and treatment changes ("Structural analysis"). Int. Orthod. **9**(4), 388–409 (2011). https://doi.org/10.1016/j.ortho.2011.09.011
72. D. Segner, Floating norms as a means to describe individual skeletal patterns. Eur. J. Orthod. **11**(3), 214–220 (1989). https://doi.org/10.1093/oxfordjournals.ejo.a035988

73. J. Lavergne, Morphogenetic classification of malocclusion as a basis for growth prediction and treatment planning. Br. J. Orthod. **9**(3), 132–145 (1982). https://doi.org/10.1179/bjo.9.3.132
74. E. Cheng, in *x + y: A Mathematician's Manifesto for Rethinking Gender* (Basic Books, 2020). Accessed: Oct. 07, 2023. [Online]. Available: https://www.amazon.com/Mathematicians-Manifesto-Rethinking-Gender/dp/1541646509
75. P. Auconi, M. Scazzocchio, E. Defraia, J.A. McNamara, L. Franchi, Forecasting craniofacial growth in individuals with class III malocclusion by computational modelling. Eur. J. Orthod. **36**(2), 207–216 (2014). https://doi.org/10.1093/ejo/cjt036
76. N.G. Strouthidis, V. Vinciotti, A.J. Tucker, S.K. Gardiner, D.P. Crabb, D.F. Garway-Heath, Structure and function in glaucoma: the relationship between a functional visual field map and an anatomic retinal map. Invest. Ophthalmol. Vis. Sci. **47**(12), 5356–5362 (2006). https://doi.org/10.1167/iovs.05-1660
77. S. Dehaene, How we learn : why brains learn better than any machine ... for now (2020). Accessed: Oct. 06, 2023. [Online]. Available: https://www.amazon.com/How-We-Learn-Brains-Machine/dp/0525559884
78. R. Milo, S. Shen-Orr, S. Itzkovitz, N. Kashtan, D. Chklovskii, U. Alon, Network motifs: simple building blocks of complex networks. Sci. **298**(5594), 824–827 (2002). https://doi.org/10.1126/science.298.5594.824
79. P. Auconi, Ortodonzia Biomimetica: soluzioni dalla natura per recidive ed eccessi terapeutici Biomimetic Orthodontics : solutions from nature against relapses. Ortognatodonzia Italiana **15**, 47–63 (2008)
80. D.J. Nicholson, The return of the organism as a fundamental explanatory concept in biology. Philos. Compass. **9**(5), 347–359 (2014). https://doi.org/10.1111/phc3.12128
81. C. Auffray, L. Nottale, Scale relativity theory and integrative systems biology: 1: founding principles and scale laws. Prog. Biophys. Mol. Biol. **97**(1), 79–114 (2008). https://doi.org/10.1016/J.PBIOMOLBIO.2007.09.002
82. J.M. Benyus, in *Biomimicry: Innovation Inspired by Nature* (Morrow, 1997). Accessed: Oct. 07, 2023. [Online]. Available: https://cmc.marmot.org/Record/.b10751920
83. D.W. McShea, R.N. Brandon, in *Biology's First Law: The Tendency for Diversity and Complexity to Increase in Evolutionary Systems* (University of Chicago Press, 2010)
84. W.L. Krieg, Early facial growth accelerations. A longitudinal study. Angle Orthod. **57**(1), 50–62 (1987). https://doi.org/10.1043/0003-3219(1987)057%3c0050:EFGA%3e2.0.CO;2
85. P.H. Buschang, J. Martins, Childhood and adolescent changes of skeletal relationships. Angle Orthod. **68**(3), 199–206; discussion 207-8, (1998). https://doi.org/10.1043/0003-3219(1998)068<0199:CAACOS>2.3.CO;2.
86. L.W. Graber, R.L. Vanarsdall, K.W.L. Vig, G.J. Huang, Orthodontics: current principles and techniques (2017). Accessed: Dec. 13, 2023. [Online]. Available: https://search.worldcat.org/title/orthodontics-current-principles-and-techniques/oclc/951917299?referer=di&ht=edition

Chapter 4
Modeling the Invisible

4.1 Principles of Organised Complexity in Orthodontics

Simplifying the orthodontic diagnosis and the treatment plan means proceeding with fundamental concepts, dividing the characteristics of the malocclusion into three components: the skeletal pattern, the face, and the dental arches. Three space planes are necessary to evaluate the harmony of the skeletal pattern. On the vertical plane, we split the face into three sectors, which must be equal in the eumorphic face: chin-subnasal, subnasal-glabella, and glabella-hairline. These simple rules attempt to neatly define certain state phases of the system, which aspects are invariant over time ('traits') and which factors are likely to vary ('states'). Patients with decreased anterior facial height and a deep overbite often have an upturned condyle. In contrast, patients with a long face have a posterior direction of condyle growth, with mandibular post-rotation, rear position of the chin, up to the open bite.

Detecting a dentofacial discrepancy means referring to the combination of several problems: anatomical disproportions, ectopic position of the glenoid fossa, obtuse angle of the skull base, resorptions, and rotations. Some simple orthodontic rules warn that when the ramus's vertical growth exceeds the alveolar processes' vertical growth, the outcome will be a short face. On the contrary, mandibular post-rotation results from excessive maxillomandibular alveolar growth compared to the vertical growth of the ramus, resulting in a long face. These simple rules of self-organisation are well-known by every orthodontist. Sometimes, however, these rules need to cover the complexity of the phenomena due to multiple interactions between contiguous and distant areas of the system. Factors external to the masticatory organ, such as postural or respiratory patterns, intervene in the pathogenetic chain.

In orthodontics, there are no "raw data": to define the anteroposterior position of the facial skeletal components, the orthodontist uses the angles SNA, SNB, and ANB to express the relationship between the maxillae and the skull base, and the degree of their mutual relationship. However, as is widely known, to define the actual information on the value of these variables, it is necessary to consider the vertical

P. Auconi et al., *Systems Orthodontics*, Understanding Complex Systems,
https://doi.org/10.1007/978-3-031-64932-5_4

and horizontal position of the S point and the N point, the steepness of the occlusal plane, and the position of the chin. The alveolar structure interacts with the roots; these determine the inclination of the crowns that interact with the soft tissues. “Full” lips can be an aesthetically appealing feature, but a “full of teeth” mouth is not, just as lips that are too thin and a profile that is too flat because of four extractions. Before proceeding with a detailed analysis of the space in the arch (inclinations, rotations, missing teeth, etc.), it is necessary to note whether any adjustments have occurred during development, whose correction will require the finding of further space. Therefore, in addition to analysing the space, it is helpful to proceed with studying the interactions between the elements of the system (a non-trivial operation).

Once the orthodontist has a diagnosis, the first decision to take is to decide the time at which to begin treatment. A quantitative retrospective analysis based on experience is brought into play to elucidate the causes of malocclusion. At the same time, a qualitative study allows one to hypothesise possible contributions from factors not fully explained. The orthodontist is inclined to plan treatment by giving therapeutic priority to the patient’s most visible problem, which, however, is not necessarily the most critical problem. Each local zone of non-random aggregation of the issues may be the cause or the effect of another problem (as is the case with anterior crossbite, atypical swallowing, and low tongue posture). As in any complex system, aggregations of local characteristics give rise to global structures, which have a feedback-constraining effect on the individual components. Once again, the local and international levels are separate in complex biological systems. The global tends to encompass and define local patterns, which at the same time are responsible for the totality. An obtuse Gonion angle conditions a global skeletal structure of maxillomandibular hyperdivergence, which in turn conditions local muscular structures, joints, postural patterns of the cephalo-dorsal-sacral spinal column, swallowing, dentoalveolar renormalisation attempts, and more. The hyperdivergent ‘whole’ allows for specific couplings and interaction between local system components, strengthening over time.

Living systems are not machines that process incoming and outgoing information by applying a function. They possess internal coherence and material consistency that can process data to generate rich morphogenetic fields in a way that machines cannot. Autocatalytic growth kinetics emerge, producing possible forms as the system onset. There are preferred states in the morphogenetic field to which the dynamic processes that generate forms are more likely to be. The orthodontist must early detect the location of the most likely stable states that make certain facial forms phenotypically improbable or impossible, knowing that the chemical-physical laws represent only the preliminary stages of the biological laws of growth. Yet, in these developmental mechanisms, physical constraints and interactions are crucial in generating variant morphological forms so that in orthodontics, the space of the imaginable features does not coincide with that of the existing ones. Theories embedded in clinical-statistical computational models measure the distance between medical theory and clinical reality, i.e., the distance between the ability to explain orthodontic phenomena at the conceptual level and the actual ability to generate data analytics and detailed predictions exploitable in a specific patient. In the following chapters, we will see how to delve deeper into these processes and localise them using computational procedures such as networks, acyclic graphs, feature selection, and others. In a biological system, the components interact to produce their behaviour patterns over time [1]. The system's structure of its internal interconnections—causes its behaviour.

Phenomena produced in the organism are not independent of the texture of matter in which they occur. In the dentofacial complex, processes occur through viscoelastic connective structures, hard or spongy bone segments, tissues, muscles, and articulations of various consistencies. Information passes through different components, consistencies, forms, and substances. The complexity of mathematical-biological matching processes makes it unlikely that clean, detailed, and repeatable signals emerge when analytical procedures develop on too heterogeneous structures, patients, and clinical scenarios. Some computational classification schemes seek to select features without removing meaningful elements to make the patient more understandable and locate the points of morphological change. The orthodontist strives to channel the characteristics of each patient into an orderly, reproducible pattern that ensures some form of regularity. Drawing on his experience, he converts features into analysable form, presuming that each component has a meaningful numerically analysable constituent, referable to general orthodontic laws. Unfortunately, orthodontic patients are too different from each other, so we cannot easily aggregate them into different collective homogeneous structures. Each patient is unprecedented, complicated, and possesses a specific kind of noise. This inherent indeterminacy is related to the simultaneous presence of conservative and evolutionary quantities, mixed with different one's adaptation to imbalance, adaptation to treatment, and tolerance.

Frequently, the sample sizes available for orthodontic epidemiological studies are not large enough to specify how even a few patients' factors (comorbidity, age, sex, level of treatment compliance, and others) could alter the benefits of a treatment or produce emergent unintended effects. Connections between dentoskeletal elements generate chains of events that reach much farther than where we have applied our therapies. Moreover, in a dynamic system, the function goes hand in hand with growth changes and compensation strategies; prognostic vagueness and treatment errors are not immediately apparent and are immediately seen and corrected. Causality is circular within the large inter-individual variability of patterns and compensatory phenomena. Causes and effects may diverge in time and space, and symptoms may appear with significant delay.

The rite of passage in all medical disciplines occurs when young, trained physicians learn about errors that older and more experienced colleagues have identified. It has been repeatedly observed that this form of "individual de-biasing" could be aided by computerised decision supports that reduce prognostic uncertainty. As in many living systems, orthodontic phenotypes vary between individuals despite constant genotypes and environment. This crucial "third source of phenotype variation", beyond form and function, is the developmental noise, in whose latent features the algorithms can extract some information.

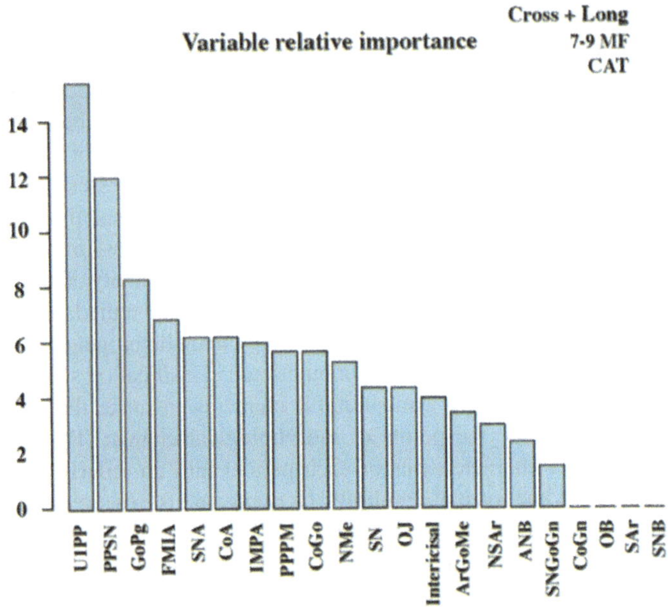
Cross + Long
7-9 MF
CAT
Variable relative importance
14
12
10
8
6
4
2
0
UIPP
PPSN
GoPg
FMIA
SNA
CoA
IMPA
PPPM
CoGo
NMe
SN
OJ
Intericisal
ArGoMe
NSAr
ANB
SNGoGn
CoGn
OB
SAr
SNB

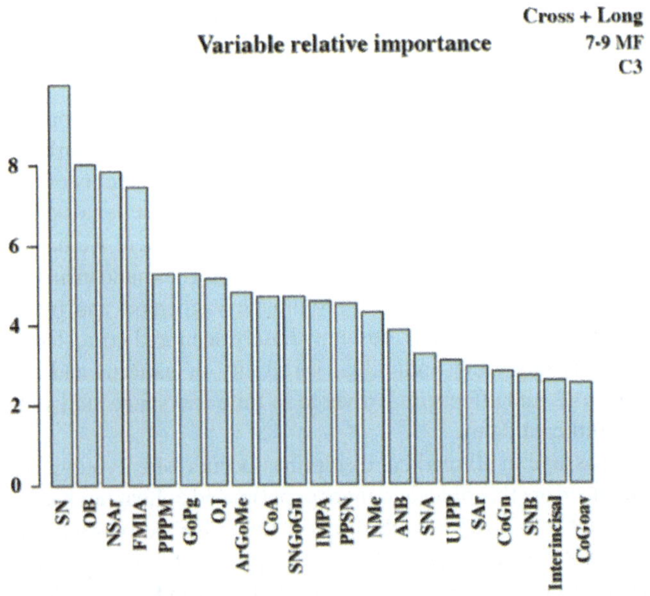
Cross + Long
7-9 MF
C3
Variable relative importance
8
6
4
2
0
SN
OB
NSAr
FMIA
PPPM
GoPg
OJ
ArGoMe
CoA
SNGoGn
IMPA
PPSN
NMe
ANB
SNA
UIPP
SAr
CoGn
SNB
Intericisal
CoGoav

◂**Fig. 4.1** Contribution (computed with linear regression) of various cephalometric quantities to the craniofacial growth of 60 patients (before the treatment) affected by Class III maxillomandibular sagittal disharmony. The Figure above shows 30 patients with severe sagittal disharmony. The Figure below shows 30 patients with mild Class III sagittal disharmony. The age of patients (males and females) ranges from 7 to 9 years. Patients with severe and mild imbalances have different essential features. In patients with mild imbalance, the contribution of features is distributed more evenly. Part of the patients used in the diagram, and in the following diagrams and Figures, come from Dr. Auconi's private archives, part from the Department of Pedodontics at the University of Rome "La Sapienza,"; part was kindly offered by Professors James A. McNamara, Jr., at University of Michigan, Ann Arbor, and Lorenzo Franchi, at Department of Orthodontics, Università degli Studi di Firenze

Random differences at the micro level can have significant ultimate morphological macro consequences. At the cellular level, the variability related to developmental noise found in genetically equal cells constitutes a crucial selective advantage by improving the possibility of rapid adaptation to new things aimed at survival processes. In plants, one can observe disparate stochastic phenomena in root configuration and different periods of efflorescence as a defence against unseasonal weather and frost. Linear statistical procedures are appropriate for systems in equilibrium, in which it makes sense to define averages. Relations emerge from correlations or by knowing the mechanisms underlying the observed properties regarding regression or causality. In orthodontics, laws governing hundreds of patients are statistical laws; laws governing individual patients are stochastic.

The scientific method took an essential first step by modelling, i.e., by defining specific quantitative characteristics that made it possible to determine simplified copies of phenomena to counter the intrinsic shaky predictability of complex systems. The defined quantities could evolve in these models according to precise mathematical laws. The study of orthodontic models attempts to enter the realm of simplified phenotypic and functional conditions to discover the mechanisms that attenuate data variation, data noise, and clinical fuzziness.

A single algorithmic reconstruction of a morphological situation, such as relative variable importance (Fig. 4.1), cannot be expected to represent more than a single part of the entire system of knowledge of all possible facial imbalances due not only to extreme phenotypic variation across the population. Feature importance is a valuable reconstruction of the importance of parameters in different imbalances, but it needs to clarify the effects of interconnection between agents in the system. Thus, this ranking does not represent the totality of the dynamics at play during facial development in contributing to the evolution of the facial complex. Some factors are not directly appreciable in the gap between the statistical description of a phenomenon and reality.

They can be measured indirectly using factor analysis, a flexible approach that enables looking for the existence of latent variables (factors) from a set of observed variables. Factor Analysis allows assigning each patient a conditional probability of residing in a multidimensional space of specific characteristics of interest. In medicine, latent feature models can search for patterns used for individual profiling, e.g., to establish the likelihood of assigning to a particular disease state. In orthodontics, latent features may favour the risk of belonging to a group of patients who will incur unfavourable facial growth and other adverse developmental circumstances such as morphological disharmonies, treatment resistance, and relapse.

In systems where there is a need to represent and simplify the complexity of a process, highlighted by many variables, orthodontists may use Factor Analysis. The process revolves around several synthetic indicators (factors in Fig. 3.8, indicated as PA1 and PA2), which are particularly significant in explaining the phenomenon. Latent factors form a giant multidimensional space in which each patient occupies a unique position corresponding to a fantastic set of characteristics. Latent features have to do with probabilities rather than specific clinical or morphological data. Factor analysis yields a commonality index, i.e., how well a visible variable couple with a particular factor, and a saturation index, i.e., how strong the relationship between the latent factor and the observed variable is (from 0 to 1; see Fig. 4.2).

Computational procedures provide a solid mathematical basis to support complexity theory and system thinking inferences. System thinking helps to understand the possible effects of the aggregate arrangements of a plurality of skeletal

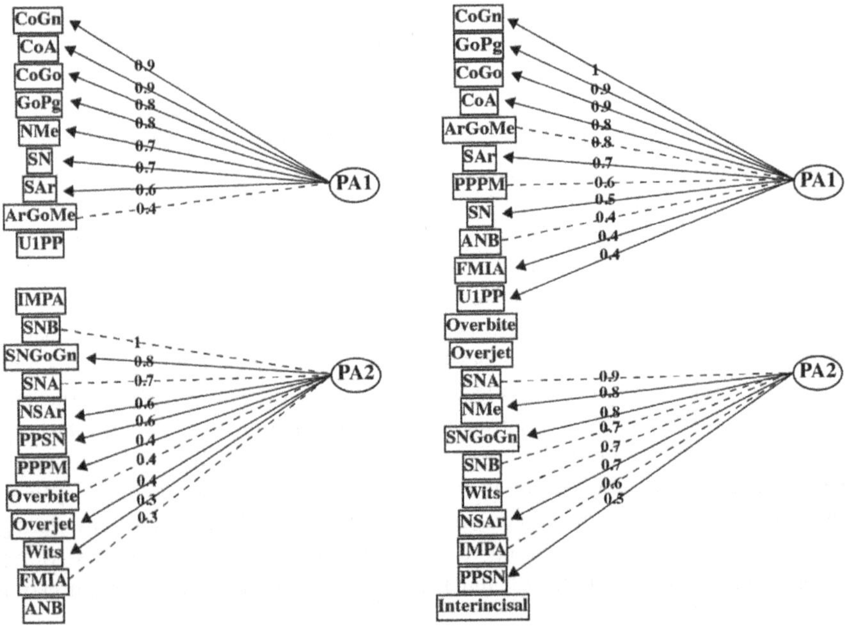

Fig. 4.2. 40 patients with Class III malocclusion (7–9 years of age) underwent orthodontic treatment with fixed devices and Delaire's mask, with favourable results (left). They were compared with 40 patients (of the same age, undergoing similar treatment) reporting treatment failure (right). The main latent factors of patients reporting treatment failure consist of 11 characteristics, the most important being the CoGn skeletal segment and, in general, factors linked to mandibular morphology. Among the successfully treated patients, bony angles appear less important in influencing outcomes; moreover, the distribution of feature strength is more uniform

interacting agents, thus determining the basic principles of evolution of these arrangements. Swarms of birds in flight provide an example of the systemic, collective wise behaviour of many agents, in which the activity of its neighbours influences the conduct of each agent. Local interactions establish privileged lines of communication that are reinforced over time, fostering the emergence of complex global behaviour. In Network analysis, the value of each element (node) is in its interactions with other nodes (links) and in its position in information redistribution circuits. Each node can connect or separate other nodes to receive information and control the dynamics of other variables, depending on the strength of the connections.

4.2 Tangible and Intangible Structures

Tangible and intangible structures characterise complex systems. Through network theory, we can observe the map of connections of the system's components in time and space. In the network topology, we can keep the factors influencing the components, pushing them in different directions over time. We can observe the system's adaptation to changing circumstances and the potential for development. By following which things are most strongly connected, we can see which parts are influencing or being influenced and whether favourable or opposing nodes surround a node. The network tells us the immediate context where we can place each patient's nodes and the most suitable soup to cook them in.

Although generated by local interactions, in biological systems, emergent properties tend to favour global characteristics and, therefore, require a broader interpretative framework. The set of nodes in a network forms an interconnected structure capable of distributed, nonlinear behaviour, characterised by repetitive behaviours and randomness. We can examine this interconnection in the facial system through specific quantification metrics, which we will discuss in detail in the next chapter. Whereas information about population characteristics stems from the aggregation of macro-data, personalised information is based on data integration in the individual patient, i.e., considering the possible relationships of feature local collaboration, competition, and reinforcement, declined in the individual patient. Anything can be helpful to understand how things caused other things in a particular patient, in a specific spatial and temporal area in which features are more important. In the exact location, the same characteristics considered in a different development period can mislead you. Accordingly, the results of many computational procedures change about periods and morphological structures (Fig. 4.3).

The shift from global to specific, individual rules is the critical point of orthodontic prediction: weaving different personal dimensions together with context sensitivity allows the extraction of additional information from the data. The facial uniqueness of each patient and its heterogeneity over time is crucial, considering the balancing processes that tend to stabilise phenotypes during the growth process and decrease variation. Disparate buffering mechanisms (canalisation, selection, regularisations)

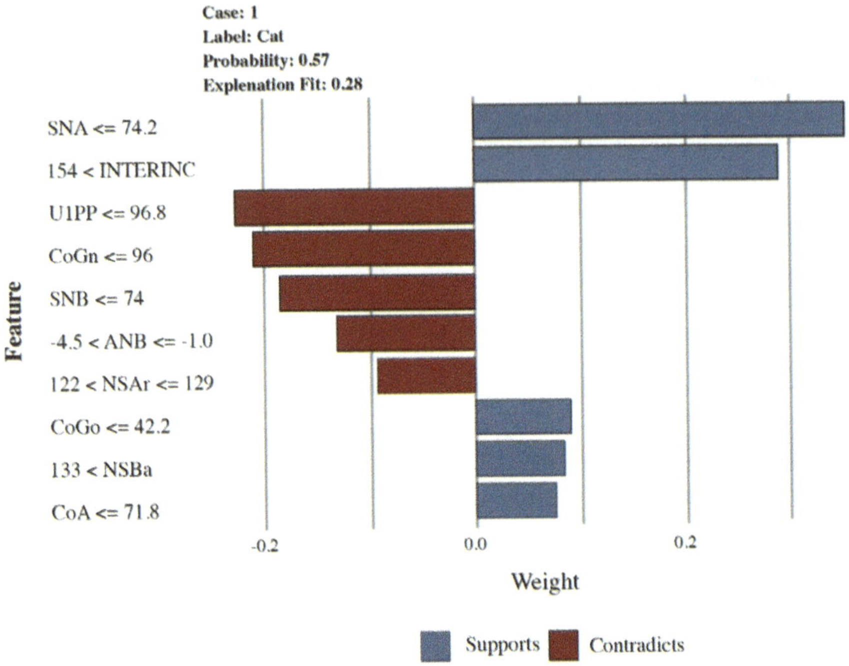

Fig. 4.3 Using the LIME algorithm, we can move from the general characteristics of a process (for instance, the onset of a disease or the nucleation of a malocclusion) to its prediction in the individual patient. LIME offers an understandable and transparent answer about the characteristics helpful in predicting a patient. Trust in individual predictions is crucial when the orthodontist uses the algorithm for treatment decisions. LIME allows us to explain the predictions of any classifier by approximating it locally with an interpretable model [2]. The Figure represents the cephalometric variables of a six-year Class III patient with severe maxillo-mandibular imbalance. The bar chart illustrates the importance of the most relevant variables that supported the early prediction of skeletal imbalance during growth. The blue bars supported the projections, whereas the red bars contradicted them

and skeletal constraints narrow the range of possible facial morphological trajectories and emergent phenomena during growth (Fig. 4.4).

The Sankey diagram shows that the prediction of the progression of a feature must recognise the constraints deriving from its morphological structure. It defines a whole natural history of craniofacial evolutionary possibilities. On the other hand, Classification Trees limit the evolutionary individual options that depend on significant turning points related to the system's bifurcation points. They also identified the threshold effects of specific critical parameters and early growth acceleration phenomena. One of the most potent growth constraints lies in the same morphological solid coherence and stability of the unbalanced severe craniofacial conformations that start early, which are capable of absorbing, at least in part, subsequent skeletal growth fluctuations. Thus, the facial system has a deterministic component in its imbalance. Beyond a certain threshold of skeletal imbalance (among patients with

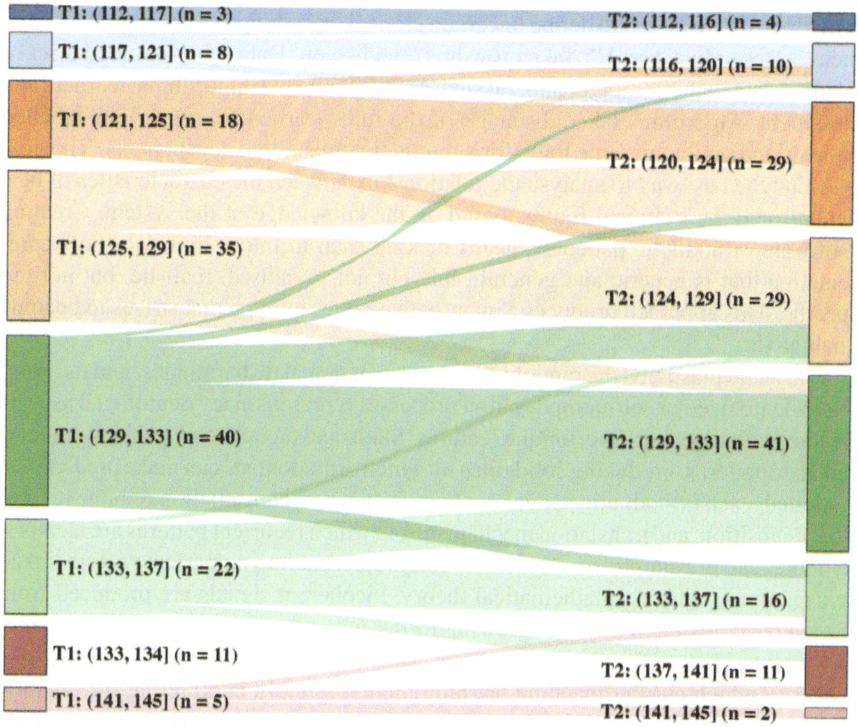

Fig. 4.4 The Sankey diagram shows the evolution of ArGoMe angle values in 142 untreated Class III patients (ages 6–17 years). Extreme ArGoMe values, high (red trajectories at the bottom) or low (blue trajectories at the top), show a quasilinear progression of values during growth. The same characteristic is present in many other cephalometric parameters of Class III patients. The evolutionary turbulence of values characterises intermediate values. Thus, the extreme skeletal imbalance is a hyper-parameter that can affect near and far parameters and distort the laws of probability—a diagram developed by Marco Scazzocchio data scientists [3]

Class III malocclusion, the severe sagittal imbalance is identifiable in subjects with Wits appraisal ← 7 mm), the morphology becomes time-bound. Structural reinforcing circuits of imbalance promote their further strength and confirm that, beyond a certain threshold, it is not the individual properties of the parts that matter most but their systemic organisation, structure, and form. When the interactions between neighbouring features are robust, the system falls into an "absorbing state," a collective behaviour that imposes a pure dominant strategy. As we shall see below, the degree of initial imbalance also affects the treatment effect.

Wanting to resort to predictive computational systems, researchers and practitioners must reproduce patient specificities by highlighting dentoskeletal differences while simultaneously fitting morphological features into a 0–1 binary system. Everything becomes data through cataloguing, labelling, and sorting into clinical categories. Design/action/expectation schemes reflect simple structures: faced with the over-abundant possible combination of factors involved for each patient; often orthodontists are forced to define uncertain clinical orthodontic situations, describing them as "pseudo–Class I," "fake crowding," and so on. Data scientists feed a set of labelled data to the computer and, on such a set, operate two algorithms: learners and classifiers. Algorithms take advantage of the rules learned from previous labelled patients and apply the same logic for subsequent unlabelled, new patients. The goal is to find the best way to analyse the relationship between the characteristics of new patients and their clinical futures based on the knowledge of the system. Creating "software orthodontic patients" means making each fit into a framework of representation that is generic and generalisable but not trivialised, realistic, but not too specific. This approach produces a quantitative representation that strives to become qualitative.

The facial plasticity, the random progression of dental disharmonies, and the enigmatic dentoalveolar self-compensation processes represent many dentofacial aspects rich in qualitative, intricate, unique features. Statistical models seek to identify recurrent data patterns, producing solid rules by which orthodontists can make predictions. As stated, each orthodontic patient contains a changing constellation of bone resorption, deposition, and translation mechanisms, in which recurrent patterns are aleatory, hypothetical, presumptive, and tricky to fully represent into numerical vectors linked to a rigorous, unifying mathematical theory. Incoherent signals are produced from the simple collision of physical growing forms and conveyed using complex biomechanics working on facial organogenesis. Thus, the fundamental issue concerns the extent to which current machine learning models, simply interpreting information flows, and reducing the phenomena of biological reality into binary codes, will be able to unravel the tangle of possibilities. We have seen that each orthodontic feature possesses a specific field of influence on other features: the features' connective intelligence lies in the logic of the organising principles. A certain amount of error is inevitable when an enormously complex world is simplified and reduced into sharp categories.

Network analysis offers the possibility of having a simplified, but not trivial, version of biological complex systems. The simplest and most natural way is to characterise them with a series of nodes, or vertices (the components of the system), and lines, or links, that join them, forming a network of relationships. This model corresponds to a mathematical structure known as a graph or network. A growing biological system seems messy, magmatic, made up of very different things, in which some components take steps forward at the expense of their neighbours. There is no reason why one relationship is established instead of another. Instead, each part carefully chooses with whom to interact. We will see that there are specific laws governing interactions (we will also see that there are robust exceptions to these laws). In the mathematical model, a relationship between vertices arises from the properties of the vertices themselves. The world's complexity has the signature of network structures where connections occur from a large 'disordered' number of probabilities. An advantage of the representation in network form is that all connections can be seen immediately by looking at the Figure. Each network shows strengths and weaknesses. The latter is often difficult to locate and can cause the entire network to collapse [4, 5].

The growing craniofacial system configures a network based on a plurality of micro agents, in which any feature becomes a node (any angle or craniometric segment one deems of interest), and the covariance relationships between nodes are the links. Each component can be considered the same way as an information processing unit. Each node links to others, but some have more connections than the average. It is not a matter of substituting concepts for insights: in many living systems, interaction patterns need repetitive subsystems to function at their best, thickenings of functions alternating with decentralisation of functions, and preferential circuits of information channelling. Unifying features testify to a widespread tendency toward spontaneous order by optimising part architecture. These connection patterns generate collective behaviours that do not descend in an obvious way from the components of the parts. The organisational network can reveal relationships of a purely geometric type (such as the relationship that binds the corners of Tweed's triangle, for example) but also unexpected functional links between specific nodes, as well as locating groupings of nodes endowed with more significant available activity (motifs), as part of a dentoalveolar renormalisation process. i.e., a subset of highly connected nodes.

When the dominant position of a particular feature of the network appears during growth, e.g., nodes having a much higher number of links than other nodes in the system's evolution, it is logical to assume that this feature, which is the system's attractor, governs the system. The attractor imposes a specific form of dominance and influences the surrounding areas more than it can be affected by them. The reconstructed topology has made it possible to highlight that the overall dentoskeletal harmony and balance are regulated by multiple attractors. Some of these forms the central assemblage of the system, and others have supporting and balancing functions (see Figs. 4.5 and 4.6).

In this situation, we must apply some filtering to extract the information present in the correlation matrix and to sort out the relevant features; for example, when we use a network representation with yEd (www, yWorks.org), the choice of filtering a cephalometric feature is usually at |rxy| > 40 to reduce the complexity of the system and to permit the identification of essential characteristics simply by visual inspection. It is easy to identify bridging nodes, i.e., nodes whose absence would split the network into two or more separate parts. Bridge nodes allow the detection of different subsystems (sets of highly correlated features) and represent the connection among such subsystems (node F in Fig. 4.5). Closeness Centrality, a measure of how much a network centres on a particular set of nodes, is a measure of how fast the diffusion of information from a given node to others in the network is. It relates to how often the node is present in the traffic load of information and indicates the relevance of a node as functionally capable of holding together communicating nodes of the system. Closeness Centrality can be calculated with Cytoscape (www.cytosc ape.org).

In the simple network shown in Fig. 4.5, some nodes appear to have more links with other nodes (nodes B, I, and L). A more accurate analysis reveals that node F is a crucial intermediary node where most system information must flow. If this node disappears, the network splits into three small networks of nodes that are no

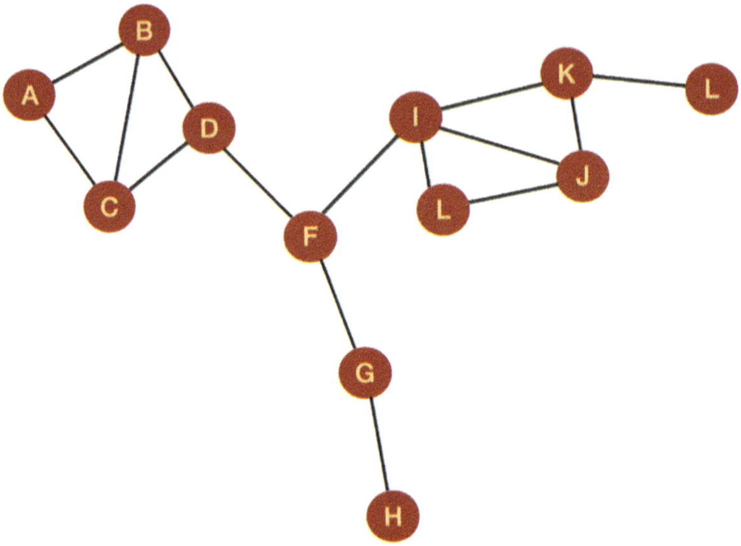

Fig. 4.5 A schematic graph plot where node F is the most central bridging two communities

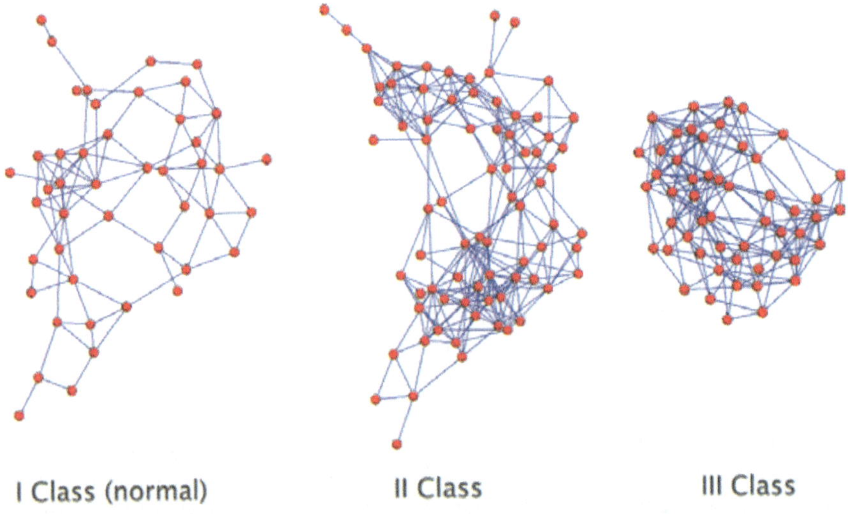

Fig. 4.6 The various networks associated with the multiple disharmonies, taken from [6]. Note the compact network structure derived from the characteristics of Class III subjects. This robust systemic structure imposes numerous constraints on the components, robust feedback systems, and behaviour over time that expose the treatment to effects that are not necessarily predictable

longer connected. The F node controls the network intermediation, or Betweenness, among the other nodes. The Betweenness of nodes in a biological network indicates, for example, in a protein-signalling network, the functional relevance of a protein in holding together communicating proteins. Betweenness is crucial to maintaining functionality and coherence of signalling mechanisms. When applied to orthodontics, network analysis deepened the relevance of dentoskeletal components influencing the growth process and unsuccessful treatment outcomes among Class III patients.

The appropriate mathematical analogy to explain the morphology of the facial apparatus is topology, the geometry of relations without magnitudes. In topology, it is possible to identify the direction of the signal or the pattern of influences of a node on other nodes. It is also possible to map the mutual influences between the cranio-maxillary components, the likely causal paths linking features, and the changes in those paths induced by the treatment (see the Bayesian graphs, Fig. 2.5 [7]).

Current tools in computational statistics based on fundamental principles of probability theory provide procedures for a better understanding and visualisation of data by learning regularities and patterns directly from the data. Bayesian statistics develops the idea of combining the information in the experimental data with prior knowledge from the literature and previous experiments to evaluate the probability of specific hypotheses. This approach is beneficial for medical practice. In Bayesian networks (also known as Directed Acyclic Graphs), links represent direct probabilistic dependencies, and arrows indicate the direction of the influence. Figure 1.5 shows the influences between the craniofacial characteristics found in 40 subjects affected by Class III malocclusion (age 6–14 years), not subjected to treatment, which had worsened the dentofacial imbalance during growth, compared with the influences found in 40 Class III subjects of comparable age who had spontaneously improved the growth imbalance. Bayesian Networks provide an intuitive qualitative description of the relationships linking craniofacial features beyond physical proximity. In subjects who improve the imbalance spontaneously during growth, the SNA node exerts an influence on the SNB, SN, and PPMP nodes, while in subjects whose imbalance worsens, the SNB node influences neighbouring nodes (SNA, PPMP). By modelling the putative causal relationships in a complex system, valuable suggestions emerge regarding treatment priorities, the location of therapeutic intervention points, and the circulation of effects in the system.

Every node in the network can be the source of some signal distortion. It may diminish, absorb, or amplify the information it receives; it may, in turn, become the transmitter, depending on its position in the network topology. The retransmitted signal is complex, consisting only partly of the received message and partly of the other impulses received by the node from neighbouring nodes. During growth, the signal energy undergoes transformations embedded in multiple feedback loops. Skeletal imbalance propagates from a generative core to related sites through differentiated material media (cortical and medullary bone tissue, joints, tendons, soft tissue, and nerve tissue). The original imbalance is a form of potential energy described as the system's propensity (details in following Chapters).

Dynamic interactions between variables take the form of mutual causal relationships. Development and shape maintenance are not effects attributable to specific variables or a "dominant" variable. They emerge from the organisation of other variables in the system. If the orthodontist refers only to physical laws to make prognostic predictions about dentofacial progression, he is dealing with an ordered, repetitive, deterministic world based on known rules. If x influences y, and y influences z, then x influences z. Such a knowable pattern persists over time: what happens next is determined by what happened previously. Orthodontic laws contain an extraordinary wealth of specific combinations that choose a rich array of growth trajectories but also a noise of unexpected final events.

Researchers define drifting averages for angles and segments during growth to frame the facial system for computational purposes. The average is a synthetic indicator of magnitude, a barycentre of the set of observations, in which no one observation should have greater importance than the others. However, the normative idea of averaging only partially fits the complexity of the dentofacial complex, which grows, resorbs, and melds itself based on a ramified interconnection of factors of different natures and different biological energy. The result of growth leads to a phenotype that combines qualitative factors influenced by highly nonlinear processes and, therefore, is difficult to define accurately in its evolution. Simple structures that interact in simple ways can lead to a wide range of complicated self-induced averages, variances, and outcomes. In addition, to calculate the average of the measured parameters, we must assign the same weight to each measurement, nullifying any evaluation of aspects of feature relative importance related to competition or collaboration phenomena or rebalancing functions between different data items.

As mentioned, the signals are pre-processed and retransmitted to other nodes (it is also possible that it happens in a distorted form). In a biological system, the persistence of the original message over time is essential: certain features have helpful 'sticky' properties that maintain the relationship between past and future. For example, features that tend to remain invariant during growth, such as the SNA angle, reflect the processes that bind the maxilla with the anterior cranial fossa and keep it constant over time. Oral breathing and deep bite are competing features, as are hypertonus of the masticatory muscles and open bite. The presence of each of these factors can heavily influence the progression of the value of another parameter, distorting the calculation of a priori probabilities. Thus, interactive dynamics is in the model as additional information. Going back from observed values to describing the phenomena that produced them requires a broad view of the occurrences. Networks can play a vital role because they give us the macroscopic map of interactions between elements and information about the local structures that determine larger-scale complexity.

A simple rule states that critical skeletal features related to unsuccessful Class III malocclusion treatment are long mandibular ramus, steep mandibular plane angle, and slight Nasion-Sella-Articular angle. Yet, not all Class III patients with these characteristics encounter disappointing treatment results. The reciprocal effects of different dysmorphic factors can be cumulative, multiplicative, distorting, and even renormalising. The Gaussian curve that determines the range of acceptability of characteristics considers globally the interconnections of phenomena. In a cohort of patients, the curve does not consider the possible abrupt discontinuities ("catastrophic bifurcations") or the almost sudden skeletal changes that may occur during the growth or treatment processes (Box 4.1).

Box 4.1. Cooperation and Balance

Combining principles derived from statistical physics and artificial intelligence, Didier Sornette hypothesised general laws and universal structures concerning catastrophic events occurring in disparate systems and circumstances, from financial markets to weather turbulence to earthquakes. These events follow statistical rules and are under the effect of 'amplifying clinical cascades' [8]. The ruinous imbalance of these systems emerges from collective processes generated by repeated interactions of local factor" A central property of such complex systems is the presence of coherent large-scale collective behaviours with a vibrant structure, resulting from the repeated nonlinear interactions among its constituents: the whole turns out to be much more than the sum of its parts" (Sornette, cit). Balancing phenomena conceal the forces capable of disrupting growth in direction and magnitude. Some local faults generate a pool of stress that ripples over neighbouring areas to the point of exceeding the capacity of normalisation phenomena. We cannot mitigate the defect through uniform distribution and dissipation over the entire system ('Constructal law'). In mechanical systems, the formation of multiple structural micro-cracks, which tend to reach a 'critical density' and result in a large fracture (a "fault"), anticipate catastrophic fractures. In the self-organisation of the facial system, the morphogenesis of malocclusion occurs from the diffusion of one or a few dysmorphic zones that connect nonlinearly with neighbouring zones, resulting in macroscopic aggregates that clustering procedures can visualise. The organisation's core has endogenous origins, controlled by the cooperativity of interaction between components.

The orthodontic practitioner tries not to be trapped in the ambivalence of data, dividing patients into separate prognostic categories by specific clinical signs, age, and gender. Applying computational systems to orthodontics, the fundamental obstacle the operator runs into is having to use a "data-driven" predictive procedure on a system where they do not know all the parameters at work.

Some orthodontic phenomena offer features with reasonably high predictability and reproducibility. Therefore, the statements above can be considered part of general orthodontic rules (as stated, albeit with several exceptions). At the same time, Table 3 reports phenomena that only sometimes follow reproducible regulations, so they have a high margin of unpredictability and, thus, a problematic margin of algorithmic representability. The general rules described in Fig. 4.1 mainly concern group tendencies, while Fig. 4.3 primarily concerns what happens in the individual subject.

Box 4.2. Make the Data Mean More

Any single craniofacial feature can have multiple representations available from different perspectives. To make a computer work, we need formal rules that lead to intelligible diagnostic procedures in the daily orthodontic practice. Sometimes, the spreading of attributes and possibilities prevents an in-depth analysis. Each feature can be subjected to additions with other features to provide further expressiveness. The Feature Engineering procedure extracts, aggregates, and refines raw patient data into formats that better represent the problem under investigation, resulting in improved model predictive accuracy. This process integrates patients' quantitative and qualitative aspects to understand and structure the biological complexity. There is a difference between representing and engineering morphological features. Morphological features can evolve, interfering with the model's long-term stability. Perceiving these subtle aspects is crucial to representing orthodontic patients successfully. The value of a variable has many possible computations rather than with the raw data, for example, the distance from the population values, normalised by age and gender. It can be related to the subject's sagittal or vertical skeletal imbalance. Each feature can be associated with its morphological counterpart. In orthodontics, we must include the effects of feature co-occurrence as important additional data in the model. Computational algorithms can "learn" in two modalities: supervised and unsupervised. Unsupervised procedures use unlabelled data to find naturally occurring groups within the data. In this first case, we may use clustering procedures to find any patterns in the data. We are still determining the result we can achieve; you try to pick up a trend in the data, grouping them according to shared characteristics. No predefined conclusions or judgments on patients' growth quality are predetermined; one lets the data speak for themselves, and the patients' details bring out the findings. Representative clusters are created based on morphological similarity ratios between patients starting from the undifferentiated data set. Obviously, in systems with many dimensions, many clusters that take various characteristics into account can be made. Clusters can reveal intrinsic patterns and hidden truths in the data.

Training with supervised procedures uses patient examples of a known output. In this case, we have a precise idea of what we need to look for, so we need the patients to be well-classified to have an accurate label. Correct classification is a powerful tool for making the right therapeutic decisions. The examples help the algorithm label the data correctly and make it more and more accurate as it works. The aim is to predict the desired outcome from novel, never-seen patients.

Regression and classification are the most widely applied procedures in supervised learning. For reliable classification procedures, a supervised dataset of no less than 30 patients for each parameter is necessary. Good sampling allows for testing the factual accuracy of the prediction. It is preferable to have the machine learn preferential examples drawn from groups with homogeneous characteristics and use local heuristics to decrease the risk of having highly accurate answers purely by chance.

In a system where everything is correlated with everything, to intervene on one part is to elicit an unspecified but substantial number of undesirable side effects. The most significant changes in the different dentofacial parameters occur between the ages of 5 and 15. However, even though about 80 per cent of boys maintain the same facial type as they grow, changes in various angular and linear parameters are not similar in the different facial areas, neither in timing, magnitude, or direction. By studying the spontaneous configurations that evolve in the networks of interacting elements of many natural phenomena, physicists have uncovered many sensitive points that can radiate information flows. The critical point is that most facial changes, whether they occur simultaneously or not, have characteristics of interdependence. Any action on this type of system can cause a branching of effects. Reducing the N-S-Ar angle brings the jaw forward. Excessive ramus growth can facilitate mandibular ante-rotation. Downward mandibular growth can ease mandibular post-rotation. Changes in chin length occur earlier in females, but the chin grows longer in males. The spheno-occipital synchondrosis can direct growth, as does the direction of the condyles, which, like the alveolar bone, are centres of rebalancing and renormalisation. After birth, functional needs (breathing, swallowing) appear and become progressively more important, causing mandibular posture to be lower and more forward and increased patency of the naso-oro-pharyngeal spaces.

As stated, it is intuitive that group skeletal trends that arise during growth are much more predictable than individual trends; it is equally evident that making forecasts based on the group trends of parameters that show little variability throughout growth would result in all-too-acceptable errors. Unfortunately, the dispersion of values is high for many facial parameters, especially those involving the chin. Therefore, prognostic predictions based on average values, as well as a priori conjectures about probabilities based on them, can have little value. Facial development depends on many variables that do not constantly interact harmoniously, temporally, and spatially. Bone structures, soft tissues (lips, cheeks, tongue), the biomechanical forces at work while chewing and swallowing, and alveolar compensation mechanisms all play a role that means that only rarely does the growth process follow a linear course (except what we have seen in the Sankey diagram, for extreme values of angles and skeletal segments). Orthodontic data analytics provide several models to integrate judgmental, statistical, and noise sources.

It is not necessarily the case that every level of macro-organisation must reconnect to the lower (local) levels of organisation. In many biological and medical systems, interactions of causality have both directions. Despite this, there is a tendency in the world of research to describe what is underlying (genes, proteins) as the cause of what is above (organs, organisms), following a reductionist agenda (which has been tremendously productive). Downward causation (from global structure to local effect) looks like a paradoxical concept but is well established as a biological fact. "...a top-down flux of causation going from an emergent level of process (say biological or mental) to the fundamental level of processes (say physical) that is supposed to underpin it, sounds like a paradox. Yet this paradox seems to be forced upon us when we want to make sense of many phenomena, ranging from psychosomatics to the mutual interaction between organismic and cellular life, or when we want to express certain facts of experience in which a change In our mental state has an immediate impact on our physiological state" [9].

A method could be to adjust forecasts if there is imprecise knowledge of the domain using the model Errors-in-variables 38. When changes in the data are not in the model, the estimates cannot be reliable. Researchers should be aware of the reciprocal inter-level cause-and-effect processes occurring in the system, at least the most relevant ones. Being able to rely on the judgments of experienced clinicians is of great help in supervised statistical responses, especially when parameter values are inaccurate or change over time. Once they have entered a critical state, complex biological systems easily pass from parameter combinatorial optimisation to combinatorial explosion, from necessary to catastrophic supercritical behaviour. During the treatment process, a small error concerning the variation of the vertical skeletal dimension can trigger various unexpected, possibly severe consequences (open bite, painful temporomandibular joint symptoms, trismus, contractures, root motility, and more). In these situations, the ability to visualise in advance the effects of feature interdependencies is crucial.

There is a computational system that can reduce potential errors concerning the multidimensionality of entangled biological processes by combining and recombining several values simultaneously, according to specific nonlinear equations: the Artificial Neural Networks. Artificial neural networks have the ability to learn patterns and associations between data without being explicitly programmed. The input data set is recoded at successive levels in order to capture hidden regularities. Each unit is only able to compute one thing. Connections can be excitatory or inhibitory, as do synapses in the brain. The higher the weight of an excitatory connection, the higher the probability that the unit receiving the signal will release. The dynamic aspect of these networks makes them capable of self-organising from random weights and connections, which gradually adapt to perform the necessary tasks.

Parallel computation of many interconnected units substitutes probability for logic. In distributed networks, each feature is distributed throughout the entire system, so that local and global processing can be combined. This ability involves 'constraint satisfying' problem solving, and problem solving in partial conflict with each other. Sets of partially overlapping features, family resemblances and incomplete patterns can also be processed following a content-based logic (as humans do). In distributed networks, each concept is represented by the global state of the network.

Conventional statistics reveal only parameters that are significant for the entire population. At the same time, Artificial Networks include parameters that may not reach significance for the whole population but are highly influential within subgroups or individual patients. So, this analysis tool meets the analyst's need to understand the world better by correctly categorising it. Categories simplify patients by reducing them into groups explicitly defined by operators or that emerge from parameters of the learning network. Since everything is given for a fee, as the Neural Network becomes more and more complex (abstract), it becomes difficult to extract the logic behind it. Traditional statistics often disregard "eccentric" patients with abnormal data, relegating them to the tail of the Gaussian, losing sight of the fact that often it is the rare case that possesses the most significant informational capacity about the dynamics of a phenomenon (as well as an improbable event that occurs). In the extreme values and patients, some authors have identified the causal primality of many facial disharmonies [10].

The fulcrum effect of maxillo-mandibular hyper-divergence from excessive extrusion of the upper molars can be absorbed and compensated for by a corresponding lower molar intrusion or a local tissue contraction. Neural Networks can represent these interactive phenomena, the nature and the strength of the interactions, and the time scale on which the interaction takes place [11]. They are a valuable tool in filtering out valid data by separating it from noise. They sift through large amounts of data before concluding. They recognise natural noise in the system, data that could be clearer or more useful for what we seek.

Computational tools can follow a further important organisational feature of natural systems: structural complementarity [12]. While in biochemistry, enzymes enormously increase the yield of functions through the specificity of molecular interactions by the meeting of complementary structures (as happens in the molecule-active site of the enzyme), in orthodontics, the complementarity of the coronal cusps-fossae sites optimises the transit of information conveyed by occlusal forces. In poorly complementary occlusal structures, such as open bite, chewing energy absorption and distribution over the occlusal plane could be more efficient. By privileging the symmetry of relations to the symmetry of dimensions, some orthodontists consider the occlusal plane the main centre of symmetry of the stomatognathic apparatus.

A complex system transforms by building structures that, in turn, govern its dynamics. Graph analysis helps to clarify the unpredictable behaviours generated by simultaneous interactions of the components at work. This approach sheds light on essential features of structural organisation, such as the integration of parts and features modularity. Moreover, it can enhance our understanding of the generation of unexpected skeletal forms during development, addressing causal and compositional relationships between facial components. The parts interact with what precedes them, and what precedes them controls the parts. In a very connected network, the behaviour of the whole regulates the functioning of the single variables, each of which contributes to the regulation of the others. A networked system condenses the incoming information in the form of some decisive data belonging to the hubs.

In a living system, the network of relationships between the components tends to assume a non-random topology, as preferential circuits emerge between some nodes, which unbalance the flow of information. The connective patterns expressed by the networks connect to the morphological organisation of the system. Anatomical parts may have greater or lesser topological relationships with neighbouring parts: distal, proximal, near or far from growth centres. When a network has few links between components, removing one node causes other nodes to disconnect. In this way, some structures can follow morphogenetic corridors independent of neighbouring structures, which may be at the root of severe morphological distortions, mainly when a variable module stays away from the system core.

The privileged hub nodes of the network, which contain many more contacts than the neighbouring nodes, are very advantageous compared to several much less frequent sites. Over time, the contact flux does not determine a random distribution according to the well-known Gaussian curve, with a symmetric mean, but to a very unbalanced curve, called the "power curve". Few large hubs contain many links, while many nodes have very few. This behaviour means that the numerous nodes with few links are connected only by passing through the large hubs, which are the crossroads of the entire system. This non-random, not very democratic structure is the organisational characteristic of many living systems, in which many minor nodes converge on a few nodes capable of ensuring the overall structural connectivity (the "small-world effect"). In this preferential attachment, we observe that the most connected nodes tend to attract more and more nodes as time passes, making the system increasingly unbalanced (the "rich get richer" effect).

The consequences and the tolerance of a treatment error will be different depending on the location of the applied force: the mistake that falls in a hub, i.e., in a critical zone full of links, will have a more significant influence and greater diffusion throughout the system (as hackers know, the Achilles' heel of the Internet does not lie in routers or links, but in hubs). Networks can help define the crucial factors that contribute to defining a phenomenon, starting from the principle that (once again) in the study of the organisation of the facial system, it is optional to analyse all the components involved in detail. Still, it is helpful to focus on the "coarse grain", the dynamic organisation of the whole, the formal rules of the personal declination of defects, and the internal collisions established between the local morphological subsystems. Network analysis helps orthodontists develop tailored plans for different situations. It considers how biological factors (like bone shape) affect each other over time, like how a child's body changes differently than an adult's. For example, an overbite might develop or worsen because of an open bite, tooth angle, swallowing habits, etc. Orthodontists analyse this complex picture using various data points and adapt their treatment plans as they learn more. However, focusing on too many minor details can be counterproductive. We can start with deductive inferences and then check whether the actual data confirms our assumptions ("Bayesian selection") [13]. As we have seen in the responses of the LIME algorithm, each subject has a spatial and temporal area in which features are most important: this inherently qualitative process is related to the interaction of meanings.

Human learning enacts several relational and associative functions in parallel: the concatenation of different planes of abstraction is made possible by the widespread distribution of processes that rest on a complex cortico-hippocampal interactive network. Circuits of facilitation and elimination of synaptic links complement this network. Synaptic distributed modulations are enacted by interconnected cells present in various brain areas, generating a concatenation of planes of increasing cognitive abstraction.

In computational systems, learning and memory abilities depend on examples. The algorithm starts from the examples to arrive at interpretation. Examples should generate unambiguous numerical features since algorithms are incapable of reasoning about what they are doing, nor do they understand why things happen. They lack structures for representing reality that resemble human concepts. Therefore, the doctor must perform a continuous exercise of accurate fact-finding. But, as we have said, many orthodontic concepts are too subtle and nuanced to be expressed in symbols that the machine can process flawlessly. For it to learn, the machine must have different stable internal states and be able to reconfigure its internal patterns by improving the functions that make these patterns increasingly useful for its desired purposes. The neurons in a neural network are interconnected, and each influences the behaviour of the others. Each neuron updates its state based on input from connected neurons, assigning a weight based on synapse intensity and activating functions to carry forward information, which will reach subsequent layers ("feedforward network").

The operator's flow of examples to the machine (for example, repeated exposure to data from patients with geomorphic and patients with dysmorphic facial features) allows the operator to modify the numbers representing synapse intensity appropriately. The goal is to minimise the error rate in recognising features and to be able to excerpt risk factors from the interconnection of multiple features. The machine improves its efficacy by performing increasingly complex computations as training proceeds with many examples. Presenting the data is crucial. As stated, each patient brings differentiated information to their clinical situation. Raw inputs are related to the corresponding label in a complicated, challenging way [3]. Individual expressivity about one's malocclusion differs from subject to subject; some patients present with a few nuanced morphological signs, others with more numerous and more severe signs. Features in the clinician's judgement that are particularly useful in defining the patient's diagnosis and prognosis ("dominant features") should be expressed and offered to the machine. As the value of a case is an additive function of its attributes, many researchers advise to multiply dominant patients by two or three times in the dataset (eventually, after minor changes in values) to train algorithms on morphological extreme pictures relevant to patient-prototypes of severe imbalance ("augmented patients") [14].

Since Artificial neural networks process numbers and not concepts, knowledge extraction comes from numerical connections, not from the concatenation of meanings. Improvement and optimisation of regression processes coincide with increasing performance speed, better selection of input features, and error compression ("error back-propagation") [15]. We need to optimise the competencies of the algorithm for specific sectors. If the algorithm learns to play chess (we cannot beat them anymore), it cannot play Tetris, even in an elementary way. In contrast, the systems optimisation mechanisms of the human mind work through refining the flexible, interactive complexity of functions distributed in various brain areas and improving the quality of cognitive abstractions. A child learns to play by teaching meanings; the machine learns probabilities. Human memory is continuously changing: the ability to concatenate planes of abstraction allows even the utilisation of noise and the indeterminacy of outcomes inherent in many real-life situations (see Box 4.3).

Box 4.3. Abstraction

While our mental models allow us to predict what is likely to happen in any situation, machines lack perceptual representations based on common sense reasoning about the world and the cause-effect relationships of phenomena. It is necessary to choose a few essential aspects of the process to learn how to measure, catalogue, and compare the wide variety of facial morphologies with each other. We must give the machine as objective a description as possible, not our reconstruction of the facts. Because, in nature, changes occur all at once, the perceptual effect of which is a succession of forms. All forms of learning require the transformation of mental representations into new ones. Abstraction and analogy are fundamental human capabilities at the core of mental models. Abstraction is the ability to recognise specific concepts and situations as requirements of a more general category. Analogy allows the ability to perceive a common essence between two things. Abstraction while learning helps form concepts that capture common properties of objects, such as relationships and hidden properties, disregarding more specific properties. It is the ability to create more general categories. Abstract concepts such as energy and force explain cognitive processes, even if they go beyond sensory experience. By seeing an apple, we can come to a broader inference; for example, we can formulate the abstract concept of fruit. Money is an abstract concept by a social convention, which allows us to exchange concrete goods for a bank bill or a credit card, that is, for something we do not observe. One can perform mathematical abstractions by extracting and isolating certain features standard to several objects, for example, three different diagnoses for three other patients and three treatments: one removes the concept of three, which unites the elements considered. Abstraction is a way of eliminating useless things. If you eliminate unnecessary details, you simplify the system and understand the situation better. In medicine and orthodontics in particular, one proceeds by empirical proofs and experiments. Material evidence is collected. Similarities between patients are sought, so that one therapeutic procedure applies to several situations. In mathematics one proceeds by demonstration, by logical concatenation of fundamental statements. It is sufficient to transform objects of a real nature into something imaginary, in which to reason logically, avoiding contradictions and making the starting assumptions clear. By converting a real clinical situation into abstract concepts, we can manipulate these concepts using logic. The advantage of using logic is that the solution is true, not 'approximately' or 'almost certainly' true. Abstraction is an ideal version of real things. An additional form of abstraction is moving from numbers to relationships.

By understanding each patient's details and using computational power to analyse various data types, orthodontists can address questions like defining facial imperfections, measuring dentoskeletal imbalances, improving critical parameters, and predicting treatment outcomes. Simplifying questions helps machines grasp them better, increasing the accuracy of answers. Natural systems utilise noise to refine self-organisation and adaptation processes. Random pairings of data streams generate information, enhancing organisational complexity and aiding adaptation to unforeseen circumstances. Adapting the algorithms to orthodontic reality, one realises that there is no predefined formalism functional for prognostic decisions, only a set of examples to structure hybrid man–machine opinions. The accuracy of diagnostic classifications is essential to obtain reliable inductive orthodontic rules. When a patient's features express an easily distinguishable facial structure, recognition is easy. However, we have already stressed that orthodontic concepts depend on context or interconnection with additional (detectable) dimensions. But some elements elude us. To feed the machine without deceiving it, the orthodontist needs to incorporate the concepts of uncertainty and contingency into the input data; ultimately, the orthodontist must become an expert in analysing missing data. A computational machine cannot formulate the concept of harmony and aesthetic appeal of a face; it defines the optimal ratios between skeletal segments and angles based on what it has learned from the data. The average of the data is judged "normal", and what statistically and acceptably deviates from the average is considered "abnormal". Disharmonies notation originates by evaluating the patient's differences from patients with the 'ideal' face, the removal from what Horowitz and Hixon called "the average expansion and change pattern".

Yet, there are general orthodontic rules that the machine must know. It needs well-specified concepts, which is a problem when defining aesthetic abstractions. Many constraints on the developing dentition are morphological and operate according to relational biomechanics. At the age of three, the relationship between the maxilla and the mandible is well established and will not change significantly after that. A rule says that a good deciduous occlusion is good insurance for an excellent occlusion in adulthood. On the other hand, defects that are already well-characterised and evident in deciduous teeth will remain in later ages. This rule depicts many aspects of craniofacial evolution and may help make predictions. Another rule speaks that, in deciduous dentition, the most favourable situation happens when (1) there is an intercuspation of Class I molar and canine; (2) there are overjet and overbite not greater than 3 mm; (3) the midlines are coincident. It is minor but may help the algorithm's inductive elementary reasoning (Box 4.4).

Box 4.4. Inductive Reasoning in Orthodontics

Patients may not closely match the typical characteristics of a condition like malocclusion. The interpretation of craniofacial measurements can vary greatly depending on the context. Standards may be flexible, concepts fluid, and boundaries changeable. Machines may struggle to adapt to these changes, affecting their ability to track patient progress accurately. When building a learning set, it's best to prioritise patients who represent the condition clearly and consistently, avoiding those with overlapping features. However, this approach can exaggerate specific scenarios, forcing doctors to decide if it's helpful or misleading. Inductive logical-mathematical processes involve making conjectures; there's no guarantee of the outcome. Clinical knowledge comes from generalising specific observations or drawing analogies. Inductive reasoning requires pre-existing theories or background knowledge. Machines need predefined concepts or heuristics; they learn solely from examples, which is insufficient for complex problems. Doctors can improve hypotheses with background knowledge. Algorithms rely only on data patterns and their context. A learning algorithm modifies its internal state by changing the strength of certain connections once we pass the activation thresholds. In any case, the decision about what is essential in the model is in the hands of the orthodontist, who must understand the premises and results of the analysis, but also the principles underlying the diagnostic process, the domain-specific constraints, the theories and assumptions that led to the results, and the concepts for evaluating competing hypotheses. In any case, the doctor must ask the right questions, using arguments related to the highest clinical forms of reasoning.

How craniofacial structures develop over time influences the importance and interactions of features, which can change unpredictably. Some variables have a significant biomechanical impact and need highlighting, while traditional math models may overlook spatial details and material interactions. Removing redundant attributes is common, but they can offer insight, especially in complex situations. Sometimes, features indirectly related to balance may be wrongly dismissed. Human learning evolves with context; we interpret new situations based on existing knowledge, continually updating our understanding. In essence, our knowledge shapes how we perceive the world. The morphology of the deciduous dentition exerts a particular influence on the permanent dentition. The study of the relationship between the terminal planes of the deciduous dentition makes it possible to establish the likelihood of developing a Class I dentition in the permanent dentition. At six years of age, only one-third of the initial occlusions of permanent teeth have the characteristics of a Class I molar setting. At 12 years, the percentage doubles, while 37% of subjects will have a Class II occlusion and 3% a Class III. Orthodontists know that a slight mesial step in deciduous dentition predisposes to a Class I final settlement. At the same time, a terminal contact with Class II characteristics invariably leads to a Class II occlusion. Most of the variation in facial patterns in the anteroposterior and vertical dimensions refer to a genetic component. Thus, there is a high correlation between skeletal features within the same family, while it is low for occlusal features: mandibular prognathism and excessive verticality of the facial skeleton have high heritability, while the heritability of overjet is almost zero. The postural activity of the tongue and cheeks is less affected by genetic influences, as muscle tone and mandibular movements influence the shape of the face and dental arches. All these notations contribute to the machine learning procedure.

References

1. D. Meadows, Thinking in systems, in *A Primer* (Sustainability Institute, Chelsea Green Pub. Co., 2008) Accessed 07 Oct 2023 [Online]. Available: https://www.iseesystems.com/store/books/thinking-in-systems/preview.pdf
2. M.T. Ribeiro, S. Singh, C. Guestrin, 'Why should I trust you?' Explaining the predictions of any classifier, in *Proceedings of the ACM SIGKDD International Conference on Knowledge Discovery and Data Mining* (ACM, New York, NY, USA 2016), pp. 1135–1144. https://doi.org/10.1145/2939672.2939778
3. P. Auconi et al., The validity of machine learning procedures in orthodontics: what is still missing? J. Pers. Med. **12**(6) (2022). https://doi.org/10.3390/jpm12060957
4. G. Caldarelli, Scale-Free Networks: Complex Webs in Nature and Technology, vol. 9780199211 (2007). https://doi.org/10.1093/acprof:oso/9780199211517.001.0001
5. R. Albert, Network inference, analysis, and modeling in systems biology. Plant Cell **19**(11), 3327–3338 (2007). https://doi.org/10.1105/tpc.107.054700
6. P. Auconi, G. Caldarelli, A. Scala, G. Ierardo, A. Polimeni, A network approach to orthodontic diagnosis. Orthod. Craniofac. Res.. Craniofac. Res. **14**(4), 189–197 (2011). https://doi.org/10.1111/j.1601-6343.2011.01523.x
7. M. Scutari, P. Auconi, G. Caldarelli, L. Franchi, Bayesian networks analysis of malocclusion data. Sci. Rep. **7**(1), 15236 (2017). https://doi.org/10.1038/s41598-017-15293-w
8. D. Sornette, Predictability of catastrophic events: material rupture, earthquakes, turbulence, financial crashes, and human birth. Proc. Natl. Acad. Sci. **99**(suppl_1), 2522–2529 (2002). https://doi.org/10.1073/pnas.022581999
9. M. Bitbol, Downward causation without foundations. Synthese **185**(2), 233–255 (2012). https://doi.org/10.1007/s11229-010-9723-5
10. G.B. Müller, S.A. Newman, *Origination of Organismal Form: Beyond the Gene in Developmental and Evolutionary Biology* (MIT Press, 2003). Accessed 13 Dec 2023 [Online]. Available: https://www.sdbonline.org/sites/archive/sdbeduca/muller.newman.book.html
11. C.T. Butts, Revisiting the foundations of network analysis. Science 325(5939), 414–416 (2009). https://doi.org/10.1126/science.1171022
12. N.D. Theise, M.C. Kafatos, Complementarity in biological systems: a complexity view. Complexity **18**(6), 11–20 (2013). https://doi.org/10.1002/cplx.21453
13. D. Morris, Bayes Theorem: A Visual Introduction for Beginnerrs. Independently Published (2016)
14. M.K. Yu, J. Ma, J. Fisher, J.F. Kreisberg, B.J. Raphael, T. Ideker, Visible machine learning for biomedicine. Cell **173**(7), 1562–1565 (2018). https://doi.org/10.1016/j.cell.2018.05.056
15. K.K.P.M. Kannangara, W. Zhou, Z. Ding, Z. Hong, Investigation of feature contribution to shield tunneling-induced settlement using Shapley additive explanations method. J. Rock Mech. Geotech. Eng. **14**(4), 1052–1063 (2022). https://doi.org/10.1016/j.jrmge.2022.01.002

Chapter 5
Phsycal Constraints

Tweed's triangle attempts to condense the geometric essence of proper dentoskeletal correlation by relating the ideal uprighting of mandibular incisors to the degree of divergence. The basic idea is that an FMIA angle between 62° and 70° ensures facial harmony and agreeableness, so any treatment must preserve it. Additional geometric features favourable to harmonic teething are as follows: in maxillomandibular neuro-divergence conditions (FMA between 21° and 29°), the FMIA angle should be about 68°; in hyperdivergence conditions (FMA > 30°), the FMIA angle should be 65°; in hypo divergence conditions (FMA $\leq$ 20°) the FMIA angle should not exceed 92°. Tweed's triangle briefly describes the concept of facial balance based on geometric compatibility, expressing the balance of dentofacial developmental processes during childhood and adolescence.

In linear systems, we can divide the problem of geometric compatibility into small pieces and add up the separate solutions. In a complex system, solutions are not addictive: the approach must confront the whole internal logic of the system with the possible presence of various kinds of distorting and forcing elements. Systems analysis, abandoning the concept of physical magnitude for the benefit of organisational pattern, does not consider clinical data from a quantitative point of view ("How far does this parameter deviate from the mean?"). Variables acquire value from other variables and to the general morphology. Clinical and radiological signs are interpreted not only in a morphological sense. They are numbers, but also (in agreement with Paul Watzlawick) "…signs for a nexus, which lack the characters of size, shape and uniqueness" [1].

P. Auconi et al., *Systems Orthodontics*, Understanding Complex Systems,
https://doi.org/10.1007/978-3-031-64932-5_5

5.1 More Lessons from Complexity

Science excels in linking effects to their causes through descriptive and predictive models. Extracting biological knowledge from data requires focusing on the right level of system description. Fundamental physics laws apply to the natural world, also in the case of a rich and complex internal structure where we must also take into account crucial biological elements. Living systems consist of countless active parts organised in a delicate, interdependent structure. Complexity theory can help harmonise general orthodontic biomechanics rules, clinical observation, empirical validation, and perceptual/predictive rules from computational procedures. The ideal orthodontic model should represent the facial system's essential aspects, nonlinearities, and nonlocalities. Are there hidden factors affecting growth and treatment outcomes? Models simplify complex behaviours, but interpreting and solving them can be challenging. Machine learning acts based on pure induction. Thus, diagnostic and prognostic orthodontic hypotheses are constrained to dentofacial features and not prior knowledge. Inductive computational analyses must deal with orthodontic scenarios full of irregularities. They must perform the difficult task of joining the dots. The orthodontic world is a dynamic space whose processes define a fluctuating level of probability rather than a rigid rule system. Each orthodontic dataset, however numerous, comprises only a portion of an enormously complex and varied feature scenario. In this scenario, unlikely events may not be rare occurrences.

Tentatively, it is possible to enrich the data by including additional feature dimensions, such as assigning the value of skeletal counterparts to each facial parameter. For example, the Gonion-Pogonion bone segment (mandibular corpus) can correlate with the SN segment (Anterior cerebral fossa) to consider additional sagittal dimensional disproportion. Each skeletal variable related to horizontal dimensional characteristics should be related to the magnitude of the GoPg/SN ratio. Variables with geometric relationships with these two bone segments must also be considered, for example, the sagittal position of the mandibular body expressed by the NSAr angle. Or the anterior inclination of the mandibular ramus, represented by the CoGoN value. However, we have correlations between each variable and the structure of the totality of variables in the system that express the same geometric characteristic.

During life, humans learn to appreciate the meaning of the world in a gradually more profound way. Computing systems do this by analysing quantitative data following a learning bell curve, which promises to establish the ever-sharper contours of trends and patterns. We know that the future is indeterminate and that we cannot instruct the machine about future experiences by inputting data from the past. We must remember that past events might and might not happen again. Therefore, the question needs to be re-proposed: to what extent can we trust digital systems that can technically not thoroughly understand the phenomenon they are analysing? Can a learning process take place without fundamental and circumstantial biological understanding? First, it is necessary to identify moments when the system is most predictable ("pockets of predictability"). Later, it is needed to predict when the system is at risk of transitioning from one growth trajectory to another (Table 5.1).

Once the clinical expressiveness of the individual features has been optimised (related to counterparts or by defining the membership to a skeletal cluster), and patients with marked characteristics of malocclusion have been privileged, we try to explain the operating statistical rule suggested by the algorithm. The empirical rules of orthodontics and the rules established by computation take their steps from somewhat distant chains of principles. In orthodontics, laws cannot precisely indicate the spatio-temporal course of phenomena; they can, however, express the limits of possibility within which growth processes occur from time to time. The operator must transfer a statistical-probabilistic datum into a biological datum.

In the space of possible rules, the predictive machine defines the boundaries of the rule and recursively searches for a numerical probability rule that matches the data. It is necessary to specify the positive examples, the negative ones, and the rules that led to the definition of the feature expected. The characteristics that determine the judgement of improvement or deterioration must be specified. The learner must meet the typical data that positively explain the concept, considering that they are stochastic variables. All procedures aim to favour convergence on clinical hypotheses that are maximally discriminative and less overlapping between clinical improvement and deterioration. Once again, the general rule is learned from examples and must be compatible with all. Rules that leave out even a few examples are not attractive. Counterexamples help to discard wrong conjectures. When the system's dimensionality is large, the number of subjects necessary to define the rule grows exponentially.

Unlike machines, humans learn new concepts without being explicitly programmed, although some acquisitions are genetically conditioned. Humans have no difficulty finding an alignment between events to form a standard relational structure. We must learn the typical pattern, which can become an abstract model that can help structure future situations [2].

These general rules mitigate the randomness present in the realm of orthodontics but, as stated, do not cancel it out. Ultimately, there is a collective effect on what happens locally. And that this joint effect exerts an influence on local factors. Much analytical evidence demonstrates that regular whole craniofacial morphologies stabilise the local systems through multiple feedback loops. Likewise, as we have repeatedly observed, very unbalanced skeletal morphologies constrain the behaviour of the components over time. Sankey diagrams, shown above, offer an intuitive insight into the binding processes on individual characteristics by the imbalance.

5.2 Noise and Reaction in Orthodontics

Orthodontics navigates between deterministic laws and biological randomness, dealing with uncertainties and unique patient histories. The progression of life events isn't equally likely due to past states influencing the present and future. Systems

Table 5.1 "Relatively general" orthodontic rules

The facial system exhibits coherent, reproducible behaviours during growth and throughout life, albeit with exceptions. Consider these rules as a PAC (Probably Approximately Correct) guide. Computational systems make probabilistic predictive inferences about final facial morphology possible: the orthodontist should also include in the scenario the level of uncertainty related to the utterances based on their knowledge and experience
(modified from: [3], *from* [4, 5] *and from* [6]
• When we achieve complete interdigitation of the cusps in a developing dentition, this will not change spontaneously over time
• The general characteristics of facial forms become fixed at the early permanent dentition
• Overbites interact with overjet. It is advisable to precede overbite correction with overjet correction
• In addition to an increase in size, there is a slight change in facial pattern, including an increase in the prominence of the nose and chin on the profile
• Orthodontic treatment has an unpredictable but generally modest effect on facial growth
• There is unpredictable variety in the pattern of maxillary and mandibular growth. Still, from the age of 12 to adulthood, the amount of the change usually will remain the same as the facial pattern markedly
• The growth changes in the mandibular dental arch between ten years and adulthood are relatively modest
• In Class II malocclusion, skeletal retroposition of the mandible causes force vectors that move the jaw forward
• Treatment must begin before age eight to influence facial growth
• There is no resorption before apex closure
• Fixed appliances increase vertical growth
• Vertical measures of skeletal pattern generally show more significant random variance than horizontal measures
• Growing patients with severe maxillomandibular hyper-divergence will incur posterior growth rotation of the mandible. In contrast, patients with severe hypo-divergence will have an anterior growth rotation of the mandible
• Physiological forces (muscle function, skeletal growth, breathing, swallowing, posture, tooth eruption, etc.) define dentition development
• If there is maxillomandibular hyperdivergence, Class II improves, Class III worsens
• Weak elevator muscles: underdeveloped coronoid process, short mandibular ramus, narrow goniac angle
• The reciprocal relationship of the bony bases influences the alveolar ratio. The latter must harmonise the constraints arising from the imbalance between the skeletal bases with the mechanical constraints arising from cusp gearing
• The discrepancy between the crowns of the teeth and the bone base also involves local compensation mechanisms
• Extractions in the lower arch should match those in the upper arch
• The treatment of light Class III malocclusion is straightforward
• In Class I and Class II malocclusion subjects, we expect average increments of facial growth to maintain (or even improve) anteroposterior skeletal relations over time

(continued)

Table 5.1 (continued)

• Avoid early tooth extractions in third-class subjects: subsequent growth will cause the reopening of the extraction spaces
• A deep curve of Spee reflects a compensation mechanism that permits teeth alignment at the expense of increased overbite
• Treating a patient with an open bite is at risk of failure (or worsening). Repeated radiographic checks are necessary
• Equal brackets and bands give different patient results, usually related to vertical skeletal changes
• Attempting to resolve severe crowding without extractions leads to worsening of labial procombination due to probable foiling of the incisors
• Early distalization of the upper first molars may result in the incarceration of the bone of the second molars
• The vertical needs of the patient influence many decisions (extraction vs. non-extraction, mechanics to use, brackets positions, etc.
• Resolving posterior dentobasal disproportion is intricate while maintaining balance and overall facial harmony
• We are putting teeth under the influence of the masseter and buccinator, which results in almost inevitable relapse
• Compare the position of the dentoalveolar arches relative to the facial skeleton during growth (reciprocal interaction)
• Posterior maxillary expansion means vertical expansion of the dentition
• Bite opening hurts aesthetics if there is a deep bite. There is a danger of resulting in a "stretched" face, with the mandibular angle opening and creating an unsightly double chin
• Correcting a Class II malocclusion with retrusion of the upper incisors makes the nose more noticeable
• Mesially inclined teeth respond better to treatment than distally inclined teeth
• The resolution of even modest anterior crowding may only be possible with the resolution of a fair and otherwise negligible Class II ratio (which the patient ignores)
• Correction of a deep bite by extruding premolars and flattening the curve of Spee results in increased occlusal force and recurrence
• To see stability after early treatment, we must determine the extent to which the intervention created new patterns, produced growth other than expected or changed angles that typically change poorly with growth

thinking analyses causal connections to understand events numerically. Breaking cephalometric data into parts and assigning individual parameters causal significance loses overall perspective and their value. Sudden changes in facial developmental trajectories may result from a "third force" between genetics and the environment, including hidden details that alter future scenarios. Irreversible outcomes can stem from lost compensation possibilities and contingencies, shaping the future. Stochastic molecular, cellular, tissue, and facial development processes contribute to morphological variation, enhancing phenotype-genotype mapping. The stochastic nature of developmental noise can produce phenotypic heterogeneities even without any other source of change [7].

New interpretative computational procedures embrace not only the probability of the emergence of a facial structural imbalance during the growth process but also the global skeletal *propensity*. They offer the chance to explore the contingency of what could have happened with the different growth alternatives. Nowadays, it is possible to monitor, at least in part, the internal state of the craniofacial system with unprecedented resolution and to follow skeletal changes related to the effect of latent factors (as we have seen in the results of the *Factor analysis*), as well as global biologic structure of *developmental noise*, i.e., the mechanism that amplifies random perturbations of skeletal features during growth. These perturbations may begin at the molecular or cellular level: damage to DNA, RNA, signal transmission networks, protein formation, and tissue organisation. These mechanisms can reach the whole facial system following local self-reinforcing cascading mechanisms. To fully understand its role in morphogenesis, the developmental noise must be separated and measured against additional facial buffering mechanisms such as morphological integration, canalisation, covariation of traits, selection, and developmental stability [8]. The anatomical features co-vary spatially over the organism and temporally over the ontogenetic time.

The naturalistic recovery of the morphology that defines the limits of the conceivable stems from the strength of self-reinforcement forces in living systems, which prepare the construction of their subsequent structure and the conditions of the stability of their development. There is a competition between developmental noise and developmental stability, mediated through mechanisms of cell signalling that regulate cell positional information and differentiation.

Corruccini and Beecher [9] observed that baboons fed a soft diet showed reduced morphologic integration between facial structures. The reduced stimulus of mastication leads to reduced integration of the facial components, leading to increased skeletal morphological variability. The stability mechanisms protect the norm of growth against disturbance by external influences. Deciphering the complex hidden structure of morphological integration would be difficult based on clinical intuition. Randomised studies to establish the real benefits of early orthodontic intervention showed a wide variation of effects. Many researchers think that subjects who responded well with functional appliances are actually "good growers" who would have demonstrated complete or partial spontaneous correction of dysmorphia over time. Some authors tend to favour the importance of lingual posture, while others emphasise muscles and patient compliance. Some think the effects of early treatment disappear when receiving comprehensive fixed-appliance treatment. Some others believe that either the genetic pattern will re-establish itself over time and that the strength of the fixed braces will dramatically alter any benefits of early treatment.

Much of the evidence in orthodontics is clinical; thus, it is intrinsically at risk for confounding factors and conflicting results. John Mew tirelessly reminded his students of the need to develop orthodontics based on logic: "...evidence can confirm or deny, but only logic can explain." In the prioritised list of addressed problems before starting treatment, we generally recommend including 'the main problem'. Since the functioning of each skeletal subsystem operates as a dynamic unit and, at least in part, is controlled by the interaction with contiguous subsystems and the overall mesio-brachial-skeletal structure, the prioritised list should ideally include, rather than the most visible defects, the most harmful interactions.

We cannot judge the goodness of our treatment decision based on the immediate results obtained. We must consider the cost of the alternatives and the possible results obtained if we had made other choices. We only observe a single realisation within a range of possible realisations. In contrast, we should be able to visualise the 'sum of the stories': what we achieved with our intervention and simultaneously all the results we did not recognise. Since randomness in orthodontics appears in the relationships of both forms and forces, the relapse sometimes observed at the end of treatment expresses the re-emergence of one or more alternative treatment histories [10]. In real life, multiple causes concur with various effects. But only some things are realised. Network theory helps discover a formal connection between facts and compare the individual clinical situation with the network of feature relationships characteristic for age, skeletal imbalance, and positioning of emergent growth attractors. An expanding device to correct a lateral crossbite in a patient with hyperdivergence may worsen hyperdivergence. The skeletal vertical dimension commands many working decisions: whatever we do in hyperdivergence situations tends to worsen the system. In Class II Division 2 malocclusion, hypodivergence resists attempts to improve vertical dimension. The interplay between morphological and developmental problems informs us about the possible trade-offs related to each issue.

In the noise of dentoskeletal variables, there is a parallel hidden biological reality that we do not see, which plays a role that we can now begin to probe using the most up-to-date computational investigative systems. These systems give us a different factual history parallel to the one we see. Instead of expectations of regularity, these algorithms are built on abstractions based on other abstractions, passing from layer to layer of computation and transforming each step. Developmental randomness must be present in the general structural pattern of processes and the related evolutionary constraints. The new whole organisational scheme goes beyond symmetry, shape, size, and structure of forms; it becomes a network of networks, a function of functions. In this way, we could analyse several organic malformation structures following orthodontic teratological laws, which are still unknown. The internal properties of these generative laws exhibit evolutionary forces and co-dependencies between anatomical parts, which tend to channel growth. Beyond a certain threshold of disequilibrium, facial morphogenetic troops are captured in a malformation propensity with unknown internal properties. In severely unbalanced patients, the function-growth-structure consecution is the opposite of the expected one [11].

The orthodontic clinical methodology is not a disembodied reasoning readily available for a cybernetic machine; the oral environment is not a cybernetic gadget, even though its problems are multidimensional and dependent on peculiar forms of organisation and component interaction. "… the tendency of large systems to evolve towards a "critical" stage in which the least interference in their state of equilibrium can cause an event of any dimension…" [12]. Quickly formulating the most appropriate equation for the circumstance, predictive algorithms exploit findings of general principles applicable in many fields. Electronic processors exhibit powerful interpretive capabilities about the hidden preconditions leading to ineffective treatments and the location of potentially pernicious dependencies among features entirely outside of the area of clinical focus. The orthodontic morphospace has revealed specific hierarchies and constraints which define relationships between components otherwise considered independent. However, something remains outside the realm of predictability (Table 5.2).

Since every morphological entity is the formal expression of content, there are inevitable conceptual gaps between statistics and orthodontics' living and reacting mobile reality. In statistics, if the correlation between A and B is 0.64, the correlation between B and A will also be 0.64. The correlation is symmetrical, but in: our case we are dealing with relations of cause-effect that are not balanced. Therefore, the study of these natural processes calls for an unprecedented form of morphological computation in which we can identify a unifying characteristic underlying the micro-dependencies of the components. It is difficult to expect a unifying rule encompassing all possible subtle specificities. The 'One Rule' algorithm can provide some insight.

A rule consists of the logical conjunction of features that indicate whether a patient belongs to a specific property. The ONE Rule algorithm aims to extract the single rule that satisfies the entire set of patients. ONE Rule is a simple classifier that generates a rule for each predictor, then chooses the datum with the lowest total error as its "one rule". *This algorithm generates one rule for each predictor in the data, then selects the data with the lowest absolute error as "One Rule."* The error is the percentage of subjects who do not belong to the rule: characteristic = patient class. It counts how often the single attribute appears among patients and selects the best patient-specific malocclusion pairs (conjunctive data). One rule algorithm constructs a single rule by consecutively adding conditions to find a local optimum in the hypothesis space defined by the data set.

In a group of girls and adolescents affected by Class III malocclusion (7–10 years of age), the parameters with the highest ONE Rule score (feature- malocclusion) were PP-PM, SNB, SNA, with an accuracy of 83.9, 80.4, and 64.3, respectively.

Living systems exhibit chains of principles that we must turn in our favour. Reforming the upper arch transversely and vertically, and then anteroposteriorly, improves orofacial motility and thus muscle tone, which in turn changes arch shape. Improvement of the transverse diameters of the palate anticipates a compensatory (modest) mandibular alveolar expansion and an overall forward progression of the mandible. Finally, progressive repositioning of the condyles in the glenoid fossae also becomes easier.

Table 5.2 Non-predictable orthodontic phenomena (before and during treatment)

• The effect of therapy on growth pattern
• Growth time and extent of growth in the individual, in specific areas of the facial system
• Occurrence and extent of post-treatment dentoskeletal defect recurrence
• Cooperation from the patient and parents

The neutral zone of dynamic equilibrium toward which we strive to arrange the dental elements should not only constitute a place of the opposition of forces but a meeting place of complementary forces, which can be mutually reinforcing. Each pair of elements is complementary, each containing something of the other. As no orthodontic situation is static, it persists in continuous flux. Because the orthodontist must deal with a multiple and fragmented reality, he must apply a nuanced, flexible clinical logic and an equally flexible computation.

We try to understand the dynamics of a system by imagining its local causes as if things could take their truth from causes. Things are implied by what they are: to elucidate the internal reasons for phenomena, we must move from causality-explanation to the logic of *implication.* According to Eastern medicine: "... the situation makes it so that it cannot be otherwise..." [13].

As each patient has its cause, we must be able to perceive his situation and evolution, configuration, and transformation simultaneously. Things constitute themselves because they evolve: the evolution of the craniofacial system is not only contained in its morphology but is at one with it. The orthodontic physician realises this if they can unravel the subtle disposition, or propensity, by moving from feature decomposition, or causal/effect decoupling, to a continuous and correlated logic of processes. They might discover how morphology is probably inclined to change in a certain predictable way to form a new whole and improbably inclined in other ways. Qualifying the "upstream" of the system, the propensity of the whole stems from situation-bearing, ever-changing factors. Dividing patients by clusters means visualising hidden structures in the data distribution, present in several relevant coherent attributes. Structural morphologies and growth facial rules express a set of antecedents that, when evaluated together, generate a consequent.

5.3 The Complementary Nature of Malocclusion

Steve Galella and co-workers stated that each face is the aggregate sum of all the many balanced and imbalanced craniofacial parts combined into a composite whole [14]. Variations in the shape of different components can produce the same overall result [15], as the living matter has polar tendencies and relationships between tendencies: reaction versus anticipation, stability versus instability, linear versus nonlinear. Different descriptions of the same morphological phenomenon are possible, depending on the declination and other expressions of competitive or collaborative (non-mutually exclusive) characteristics. The opposing tendencies may coexist simultaneously: in some individuals, the propensity toward polarisation may prevail over integration. These are the subjects in whom we recognise many of the typical dentoskeletal features of a specific malocclusion. In subjects with more nuanced morphological features, we detect a tendency to integrate different dysmorphic aspects.

Polar morphological features represent ideal states of concepts regarding malocclusions to the highest clinical expressiveness; the reality of the orthodontic clinic lives in between these poles. We have seen that, through Fuzzy logic analysis, we will see the possibility of assigning a membership of each patient to a malocclusion by determining the degree of belonging to a given malocclusion and, for each malocclusion, the level of sharing of typical characteristics. A high membership in a specific malocclusion tends to make some characteristics permanent and to maintain the global characteristics of the facial skeleton constant. A very pronounced Spee curve attracts the hyper-eruption of the upper premolars (or vice versa). From these tendencies toward the polarisation or integration of morphologies, we can indicate whether a given growth pattern will likely remain or leave its cluster. Let's move beyond the symbolic approach of dentoskeletal apparatus, based on steady states and fixed points, to consider complementary aspects and coexisting trends such as segregation-integration and competition-collaboration between elements. We can obtain indications of a system's tendency to persist over time.

Moreover, we can clarify what the actual degree of connection is between statistical prediction, morphogenesis, and positional information and make the appropriate adjustments. Researchers of complex systems have observed characteristics of these systems on an 'extended time scale' or the effect of a defined local treatment on the whole system. Variables that persist over time are more influential and significantly impact the system's final structure. These dynamics are essential when formulating the treatment plan, considering skeletal variables continuing over time, which are particularly valuable and probably more reluctant to follow our intentions. Time-structured observations help explain causal relations between features we cannot digitally simulate or experimentally manipulate. Longitudinal models are crucial to deepen notions of change and stability as intra-individual phenomena (i.e., within-patient phenomena) [16]. *Traits* can be regarded as stable and permanent attributes, while by *states*, we mean floating attributes dependent upon a temporary constellation of influences.

It is necessary to differentiate systematically the conceptions of stability and change when longitudinal data are available. Covariance stability is often translated into a regression of a variable on itself in longitudinal data. Only longitudinal data express a patient's life change, while cross-sectional data provide direct evidence of a feature's stability, or lability, in a group of patients. Traits are potentially helpful to discriminate between individuals and are valuable predictors of other attributes. A variable can influence itself: the *autoregressive effect* determines that prior values of the same variable can cause values of a variable. Cross-sectional studies describe relations at only one point in time. As causal effects take time to occur, no cross-sectional model can be an appropriate model of a causal relationship. However, when discussing semi-supervised learning, we saw the usefulness of background and cross-sectional data in enhancing patient information over time. Often, models fitting longitudinal data need to consider time lag properly because they ignore the effects a variable can have on itself [17]. Moreover, "...effect sizes can vary as a function of the *length* of the time lag between a cause and the time for which its effect is assessed" (ibid.). Unlike cross-sectional models, latent longitudinal models can make explicit assumptions about time lags and prior values of features.

The individual patient represents the "coordinating unit" of the possible tendencies. We can quantify the degree of morphological integration by the intensity of the statistical association in a particular phenotype, i.e., the high membership to a cluster. Importantly, fuzzy cluster analysis also indicates each patient's degree of belonging to other malocclusions. In other words, a patient with Class II malocclusion may have some features typical of Class I malocclusion. So-called 'minority' characters may gain importance over time and become predominant. Thus, as we have seen about selecting less critical data, any data can become significant by changing circumstances.

Every forecast involves a systematic component related to the information available during the forecast and an unsystematic component related to several random factors, from measurement errors to operator inattention. The greater the unsystematic components of the prediction, the lower the accuracy of the response. Any test score on which to base the patient's future facial appearance prediction is assumed to be a sum of an actual score plus an error. Predictive procedures based on observations of specific dentoskeletal features are more than complete and connected to the underlying whole pattern. They also require idealised circumstances ("all else being constant", "*ceteris paribus*", *etc.*), which do not consider situations of systems constantly changing and adapting. The orthodontist might guess that a rule valid for a given patient is also suitable for another patient who resembles him. As we have seen, it can confirm the validity of the hypothesis using the findings of algorithm *Case-based reasoning*. Orthodontists have yet to constantly update news of changes occurring in the lives of their patients. These circumstances can affect the variables and the forecast outcome [18].

Before submitting a specific question to the machine, data can be revised, supplemented or corrected (i) if there is a high degree of uncertainty in the request for information, (ii) when there are changes in the patient's history, (iii) when there are unexplained profound changes in the accuracy of the response in different patient groups.

The prediction cannot be accurate when essential patient history changes are absent in the model. Statistical procedures such as Factor Analysis and Bayesian networks may lead to the suspicion that latent variables may have played an essential role in the final data. Moreover, it is possible to define the specific contextual information and locate the actual dominant variables of the system. Partitioning the multidimensional state space into specific dysmorphic regions offers a prognostic space for evaluating the progression of the malocclusion [19]. The cognitive path of orthodontic diagnosis consists of continuous disjunction, reduction, and forced categorisation. Clinical discomfort originates from the overlap between stable diagnostic-therapeutic schemes, the vagueness of biological diversity, and event temporality, whose misunderstanding is often at the root of the phenomenon of 'overtreatment'.

When we encounter a patient with a Class I malocclusion, we know there is no imbalanced relationship of the bony bases: the discrepancy is found, possibly in the vertical and transverse dental relationships. Crowding is the most common feature of Class I malocclusion due to dental-basal disproportions or loss of space related to genetics, early tooth loss, carious processes occurring in deciduous dentition, irregularity of the permutation, or trauma. Beyond the laws of biomechanics, organisational laws can radically change the relationships between the system's constituents, making us reconsider the initial principles of treatment in depth.

As a system becomes more complex, it becomes more organised. This dynamic is common to all living systems: the components optimise to coordinate movements. In skeletal morphogenesis, elements complement each other, showing that each part reflects the whole, and the whole contains parts. On the whole, there is a part. In the part, there is the whole, to a certain extent (Kosko 1993). The quality of the dental exchange also plays a role in confusing prediction rules. The excellent quality of the dental exchange corresponds to the regular and harmonious succession of events that, within 6–7 years, lead from deciduous to definitive dentition. The dental elements interact with each other strategically. Often, making the first move and anticipating the moves of the antagonist means winning the game. The most favourable eruptive sequence of permanent teeth, the one most frequently associated with the best Class I occlusion, is listed in Table 5.3.

We have already seen that, in orthodontics, the concept of normality corresponds with what is acceptable rather than ideal. An example is the wide range of definitive dental elements eruption times. Mixed dentition announces with the eruption of the first definitive molars and the lower central incisors; the latter are usually more lingual than their predecessors. A series of adaptive phenomena resolves the momentary dento-alveolar disproportions that occur during mixed dentition. Evaluating the congruity of the total arch space concerning coronal diameters reassures that the incisors can gain a physiological position with the eruption of the canines. It is easy to observe some degree of crowding and rotation of the crowns of the mandibular incisors after the eruption: if modest in magnitude (2–3 mm), the disproportion will resolve in parallel with the natural phenomena of arch settlement and growth. The intercanine distance increases by 1.5 mm at the lower lateral incisors eruption. Still, crowding is unlikely to resolve spontaneously with more substantial dental-basal disproportion growth.

In the upper arch, the natural settling strategy put in place by the permanent incisors at the time of eruption is to choose a more proclined direction to rely on a larger alveolar perimeter. If the dentobasal disproportion is too large, the lateral incisors will have no choice but to erupt in a palatal position, often resulting in a crossbite and a form of mandibular dislocation in closure, situations that will tend to persist over time. The upper canines develop and erupt from a palatal position, heading toward the vestibule behind the roots of the lateral incisors. It is essential to ensure that the

Table 5.3 Most favourable eruptive progression

Maxillary upper jaw = permanent first molar, central incisor, lateral incisor, first premolar, second premolar, canine, second molar
Mandible = permanent first molar, central incisor, lateral incisor, canine, first premolar, second premolar, second molar

canines are palpable in the vestibule by the age of 8–9 years. If the deciduous molars remain in the archway until the period of physiological permutation, we can count on lee-way space, an additional 1.5 mm (maxillary) or 2 mm (mandibular). At the time of mixed dentition, a further increase of about three mm in the upper intercanine distance is possible (before the age of nine); a further minimal rise until age 13 is also likely. The width between the buccal cusps of the second premolars tends to increase from age 3–18 years, but only slightly: 2–3 mm. In the lower arch, the circumference tends to *decrease* by about 4 mm, parallel to the anterior unwinding of the definitive molars. In mouths that tend to crowd early, the loss of space is even more significant.

Ultimately, the quality of self-organising mechanisms during the permutation is related to a delicate succession of different circumstances: an increase in inter-canine and intermolar diameters, resorption of the anterior margin of the mandibular ramus to allow for the accommodation of second and third molars, dentoalveolar vertical growth sufficient to maintain an excellent occlusal relationship. The orthodontist intervening early in this magmatic dentoskeletal reality, with heterogeneous features still becoming, calculates the amount and direction of the treatment forces to be applied. However, he needs help precisely calculating the zones of natural absorption of these forces, the radiation of force vectors, or the direction of the corridors of force disposal. The biomechanics of therapy might oppose one imbalance component of a patient to favour another. The mutual interaction of the various components will undoubtedly be affected, as will the *overall propensity* of the developing facial system.

The application of Fuzzy cluster analysis to a group of male and female patients affected by Class III malocclusion (6–18 years of age) allowed to characterise three distinguishable phenotypic clusters: subjects with increased mandibular sagittal dimensions, subjects with increased maxillomandibular divergence, and subjects with more balanced, intermediate characteristics between the first two. Respectively named are HM, HD, and Bal patients. Moreover, the fuzzy clustering made it possible to assign each patient the exact degree of membership to each phenotype. Most importantly, the subjects who, at an early age, exhibited a high membership grade to one of the three "winner clusters" (over 60%) *remained in the same cluster during the growth process* [20].

This feature indicates the close relationship between robust morphological structure and growth characteristics. The growth process is a competition between schemes; it depends upon the origin size and the spatial relationships between the different facial components [21, 22]. Foreseeing means glimpsing unfavourable factors when they have yet to materialise. More than fifty years ago, Horowitz and Hyxon stated that the size of the facial dimensions at one age was of minimum usefulness when used as an indicator of future growth. No morphologic trait could enhance the orthodontist's ability to predict future changes due to the comprehensive individual variations from the average [4]. Other researchers have found weak or fragmentary signals indicative of facial developmental characteristics [23, 24]. Fuzzy clustering allows the possibility of mitigating the variability within individual growth trajectories in Class III subjects, assigning the membership percentage to each patient to a specific whole craniofacial phenotype. We realise fuzzy analysis through the evolutionary strategies and the current convergence of these strategies in distinct individual phenotypes. The area in which craniofacial changes may happen is small and limited by the nature of the changes that have occurred previously [25, 26].

In the recursive dynamics of living morphologies, effects are the causes and producers of what generates them. Therefore, we must be able to visualise things simultaneously in their morphological configuration and transformation. That is, to see the process in the form. Once the extremes of the patient's evolutionary virtuality are known, we must grasp the process's propensity *and even certainty over time.* Instead of the conditioned reflex that leads us to deconstruct what we see, let us strive to trace the internal coherence of processes, the factors that have operated to give stability to the situation, and the network of compensations and propensities. We must focus on "what is about to happen" based on "what has just happened" because one implies the other. Suppose events are contained in the structure, indicated by what they are. In that case, the perception of the actual structure hidden in the data allows one to appreciate the future mutation already at work in the present structure in terms of *propensity* and *unfolding*. The unfolding predicts in advance what will result and, by that, corrects continuously just enough to reorient through the sheer propensity of the forces at play. As repeatedly stated, a patient's evolution is not independent of his whole morphology in systems thinking. Propensity allows us to subtract happenings from accidental events, visualising how much we can predict the future, given what we have observed in the present. The errors we find in the linear forecasting approaches, such as linear regression, are the deviations in the function linking propensity to unfolding between model and individual propensity.

In a situation of complex dental permutation, living matter can deploy winning propensity strategies to accomplish the best unfolding, including taking advantage of various dental accommodation systems. Under insufficient space, lateral movement of the incisors and shifting of the lower midline may facilitate the eruption of contralateral canines. The natural compensatory factors put in place during the permutation (mesial drift, rotations, coronal migrations, lee-way space exploitation, and so on) are "load-bearing factors." (Fig. 2.3).

According to Francois Jullien, Western medicine, having a general ideal model, strives to make the model fit the facts [27]. In Oriental medicine, a factor is loading bearing when it is spontaneously devoted to a specific development, which we can leverage. Instead of making everything depend on our initiative during the orthodontic treatment, we recognise that a particular potential is present in the morphological situation, which we must locate and be "carried" by. "...we must privilege a conception of efficacy that teaches us to let the effect happen: not to target it (directly) but to imply it (as a consequence); that is, not to seek it, but to gather it, to let it result. It is enough to know how to draw party from the developments of the situation..." [13].

In the skeletal-geometric data of the patient, we must try to identify the virtuality that distinguishes favourable or unfavourable development, the potential already contained in the early data. The orthodontic strategist's motivation must start from the actual situation, not the model. Treatment must follow the internal logic of factual development, not the logic of wanting to fit facts into the model at all costs. The gaps between the regressive model and the real one must tense the thoughts, setting conceptual gaps in motion. Visualising the patient's situation and evolution at the same time allows the distinction between assumptions and practice to be blurred. Evolution is one with configuration. Network science teaches that the events are in the structure, in the metrics of the system topology. Instead of searching for the causal/caused breakdown of the regression algorithm, it must switch from a logic of explication to the logic of *implication* of the internal reason for the process. As each moment contains the next, the treatment must not disrupt the established facial morphological order in a lacerating manner, provoking an opposite reaction. True transformation does not arouse resistance; it moves things into their natural unfolding. This non-separation of thought must guide the corrective intention, the subtlety of a flexible strategy.

Our treatment action is local and temporary; the fundamental transformation must be global and progressive. Computational procedures do not elaborate simply correlated data; they move bodies where each feature contains its evolutionary succession. Thus, the 'initial conditions' to which complex systems are sensitive and the operator must grasp should have morphological and process findings. The components of this disposition are already present from the beginning and, as defined by the Tree responses, can be identified as specific risk factors for unfavourable growth [28].

Sometimes, to implement this disposition, it is necessary to distance ourselves from the thought from which we come. Instead of aiming for an ideal model form (the implementation of which always involves a certain amount of forcing), take advantage of the 'leading' values of the situation and let them drive the action. Francois Jullien says, "In Europe, modelling thinking has led to thinking of effectiveness in terms of means-ends. First, there is the ideal plan, and then the most appropriate means are found. [.....] reality is, in fact, a succession of transformations ...the action is salient, to effect, conspicuous, but it is only the transformation that is effective. Action, at best, can be a focus contraction in the continuous course of things. This structurally condemns it to remain ephemeral and superficial" [13].

During morphogenetic processes, each cell carries a complete copy of the genome, which reads only a small portion, to receive information about further development and how it assembles with other cells to form cell "communities". Much information comes locally, directly from the surrounding cells. Each cell behaves because of the behaviour of neighbouring cells through a complex network of chemical messages exchanged through intercellular junctions. At the macroscopic level, the morphogenesis of the facial skeleton exhibits a variety of repeatedly used structural balancing mechanisms of internal parts, multiply localised: the definition of dependencies among local skeletal components makes it possible to highlight global deviations, fault circuits, and hidden side issues, related to the most imperceptible patient specificities.

Mereology is the formal study of the part-whole relation, constituting ontogenetic research's logical core. The orthodontist's challenge is deciphering the complex biomechanics of the signal network and predicting the emerging global behaviour from local interactions. We can think of adaptive behaviours such as tooth crowding, which may attempt local compaction of occlusal forces, and incisor proclination as an anterior structural failure to compensate for posterior occlusal imbalances. An increased curve of Spee as an attempt to optimise occlusal contact, a lateral crossbite to rebalance a postural asymmetry, and so on. We can imagine these aspects arise from reactive underlying biomechanics of orthodontic phenomena.

Algorithms can identify areas of defect nucleation (*hubs* of the Network analysis), the direction of spread of disharmonies (*Shortest paths* of the network), the effects of mutual regulatory influences between features (*Bayesian Graphs*), the nesting of fulcrums of inertia (the *selection* of unimportant features), the location of specific risk factors (paths and thresholds of *Regression* and *Classification Trees*), latent morphogenetic agents (*Factor* analysis), specific growth constraints in severely imbalanced patients (*Case-based reasoning*), location of growth attractors *(modules)*.

An important issue concerns the comprehensibility of Machine Learning responses and the relative ability to trust them when making essential treatment choices, such as dental extractions. Orthodontic practitioners should be warned of the related benefits, shortcomings, and risks, as algorithms that learn from human decisions likewise learn human mistakes and biases to have confidence in these procedures.

We begin to glimpse rules, regularities, effects, and the hidden structure binding data together so far that we could not probe using simple statistical procedures. Due to the sparsity, noise, and time-dependency of orthodontic data, tools of Artificial Intelligence are raising unexpected problems related to incorrect answers related to the mismatch between human clinical reasoning and the outputs of algorithms. The orthodontic discipline is working to mitigate these issues by developing more representative versions of the orthodontic phenomena and more clever computational metrics. We will see the progress made and the work still to be done.

5.4 A Matter of Trust

Research to clarify the Inner logical-mathematical processes of data mining algorithms in providing answers has yet to lead to consistent progress. The problem mainly concerns the operational 'black box' of Artificial Neural networks. Examining the mathematical structure needs to give more insight into its advancement steps. Even if, in recent times, the Neural Networks are releasing some information on their internal structures (thanks to the precious work of some research groups), the rationale behind many digital answers remains largely obscure. In the medico-legal field, diagnostic errors pose significant challenges for assigning responsibility. Operators must merge inputs into unified processes to address vague computational responses and diverse data. Since measuring internal biological states is often tricky, operators rely on experience and intuition to prioritise specific risk factors, adjusting linear regression weights or deferring decisions to "Deep" recursive algorithms. Orthodontists must creatively interpret raw data to enhance each feature's significance by leveraging data relationships and patients' confidence levels in the diagnosis. Belonging to a morphological cluster can also help in this case [29].

Orthodontists have long felt the need for a tool to synthesise the large amount of data available for each patient into operational decisions. Machines are not simply passive receptors of information; they actively learn to interpret complex orthodontic phenomena, clarifying the general regularities and the adaptive solutions at work in the individual patient. Morphogenesis deals with the succession of discontinuous skeletal changes, and their origin arises from small, continuous changes. To decipher the path leading to the result of a biological process, that is, a coherent and stable craniofacial structure, the orthodontic practitioner must delve into the general regularities of system behaviour, starting from the micro-phenomena of local skeletal determinisms. Malocclusion can create unfavourable occlusal guidance for balanced growth, causing an ultimately defective occlusion, making dentoalveolar compensation mechanisms ineffective and the regulation of local processes difficult. If a blocking occlusal guide occurs, it can worsen maxillo-mandibular relationships. In other words, a conflict between local adaptive solutions and global occlusal balance strategies may happen in the generative process of malocclusion. Computational machines must be able to learn these processes as well to provide reliable predictive answers about the treatment outcome,

Skeletal processes are dynamic, while many digital methodologies are snapshots of the morphological state at a given time. This property may provide a skewed image of the actual orthodontic situation. Problems are often treated as steady-state processes, while orthodontic processes are historical. Many years have passed since Charles Tweed stated in 1945 that four dental extractions were required in more than 50% of patients to achieve functional and aesthetically satisfactory dentition. Today, we trust in the alternative way of being able to influence facial growth by camouflaging imperfections based on more advanced therapeutic resources, the better understanding of the laws of biomechanics, the proper exploitation of local periodontal reactions, the establishment of the ideal condylar position, and so on [30]. We resort to extractions or maxillofacial surgery in a selected case when you can't help it, anticipating treatment outcomes with visual simulations.

The tolerant approach to the possible characteristics of the masticatory organ also stems from the emergence of the idea that dental extractions constitute a form of depletion of the dentofacial system, which sooner or later will present its bill. In growing systems, the emergence of new qualitative capabilities requires achieving a critical mass of factors at play. As Kevin Kelly notes, "…one grain of sand cannot produce a landslide, but put enough grains of sand together, and you get a dune that can kick off a landslide. Certain physical attributes, such as temperature, depend on collective behaviour. A single molecule floating in space does not have a temperature. Temperature is a collective characteristic of a (huge) group of molecules. Science has long acquired that large numbers behave differently from small ones. Crowds generate the required measure of complexity for emergent entities. The total number of possible interactions between two or more members increases exponentially as the number of members increases. At a high level of connectivity and with many members, crowd dynamics take control. 'More' is different" [31].

The clinician called to manage the orthodontic "more" may follow the internal logic typical of a mechanical system: he may decide to lighten the ballast of the crown-radicular biomass, i.e., perform extractions. Intervening with this collective entity of "distributed essence" by operating a subtraction of "capacity" may have depressive effects on the system's skeletal adaptation, growth, and function resources. Alternatively, he can follow a more subtle logic based on the acquisitions resulting from a mixture of effectors and parallel functions. But there is still the risk of making a wrong decision (see Tables 5.4, 5.5, 5.6).

In Table 5.6, we see that following the logic of distributed systems does not provide only advantages. The logic governing dentofacial development can be unpredictable and risky. It involves combining reactive and functional elements, where problems sometimes accumulate in more straightforward ways. The laws governing the entire system differ from those governing individual parts. There are more potential combinations of local disorder than ordered functionality. By age twelve, more individuals exhibit some form of dentofacial irregularity than those with perfect dental alignment. Nature optimises growth by creating symmetrical shapes and seemingly irregular structures, some of which may appear unsightly. Symmetrical structures maintain dynamic equilibrium more efficiently, requiring less energy to function effectively.

The proprioceptors continuously send an enormous amount of information to the central nervous system about possible "defect states" of a system. The nervous system sends orders to the muscles (temporary adaptation, with significant energy expenditure) and from them to the connective tissue (definitive adaptation, with energy optimisation). All this information undergoes a continuous process of self-monitoring. The transition from deciduous to standard dentition is characterised by increased surrounding musculoskeletal biomass and several connections and information embedded in the system. Each dysmorphia contains, in the three planes of

Table 5.4 Characteristics of distributed systems

• An absence of imposed centralised control
• Autonomous nature of subunits
• High connectivity between subunits
• Nonlinear spiderweb causality of peers influencing other peers

Modified from: Kelly [31]

Table 5.5 Advantages of distributed systems

• Evolvable—systems can quickly move places of adaptation from one part of the system to another
• Capable of recovery—there is an overabundance of systems in parallel. Small failures disappear in the system
• Borderless—positive feedback can create new scaffolding
• Generating novelty—the measure of the effect is not proportional to the measure of the cause. There are infinite new possibilities in the exponential combinations of many interconnected factors. Individual imperfections are allowed

Modified from: Kelly [31]

Table 5.6 Disadvantages of distributed systems

• Suboptimal distribution of resources—duplication of effort abounds due to redundancy of functions
• Not controllable—there is no command centre
• Not predictable—an emerging novelty can also be catastrophic
• Not understandable—systems with self-produced logic, sometimes paradoxical. Intersecting logic, A causes B, B causes A. Responsibility for the actual cause is spread horizontally throughout the web. The trigger for a particular event may become unknowable
• Not immediate—the diffuse system needs time to kick in. Lateral causes need to stir around and then come back to rest

Modified from: Kelly [31]

space, the mutual contrastive influence of tensor forces and the set of eumorphic evolutions expressed only partially. In fact, despite the large number of variables and the crossing of numerous unstable situations to the final dentition, only a few occlusal arrangements are realised.

In the daily work of orthodontic operators, the risk of considering random historical facts as a valid basis for learning is high. As is intuitive, the more forecasts cover an extended period, the less accurate they are. Weather forecasts for longer than a week must be more precise in climatology. However, suppose the morphological memory of the facial system at the initial conditions is strong (such as in the case of characteristics of severe maxillomandibular imbalance). In that case, some background characteristics remain for a long time. Much of what we know is uncertain or provisional. There are data that we do not see, which can have a significant effect on our prognoses. It is not only the cooperative or adversarial synergies between the components at work that muddy the waters.

Paraphrasing David Hand, we can call this data that we do not see 'dark data' [32]. These data need more visibility and numerical weight, but when framed correctly, they can prove very dangerous in distorting our conclusions. Things seem different because there was too much unseen data, the passage of time has obscured valuable meanings, we needed to update our assumptions promptly, or we needed to think more about what might have happened if we had acted differently.

Computer operators guide machines by specifying specific parameters, helping narrow potential hypotheses. Without predefined hypotheses, computer systems would need to start learning from scratch. Simplifying models allows knowledge gained in one area of the system to be applied more broadly, aiding in understanding other parts. Despite ongoing improvements, it's the operator's responsibility to help the machine utilise data effectively. Similar to how infants quickly learn to recognise faces and children learn how balls bounce, individuals use practice and inherited assumptions to optimise their actions. They extract essential insights into how systems function by adjusting the probability of success for each action. Children effortlessly integrate new information into their existing knowledge network, uncovering general rules underlying specific cases. The craniofacial complex is intricate and can surprise clinicians with unexpected developmental paths, leading to unforeseen occlusal patterns and treatment outcomes. Some occlusal issues may only emerge during late adolescence or early adulthood. Everyone's mouth is unique, creating unique challenges. Persistent growth distortions may result from feedback loops or disharmonious disposal mechanisms, leading to malocclusions. Complex systems organise themselves through hierarchies, which help respond to external changes and reduce the need for excessive memory. However, focusing too much on local goals can lead to suboptimal morphology. During growth, the system tends to follow paths that minimise energy, dispose of entropy, distribute masticatory forces evenly, arrange occlusal points symmetrically, and position the temporomandibular joint optimally. The system minimises errors to prevent lesions, asymmetries, and functional discomfort. In natural ecosystems, there are no proportions and regularities subtracted in advance from local contingencies, potentialities, events, and coherences "... order is not improbable, then, in nature: exchange with the outside makes it possible and, in a way, imbued with disorder. It arises from exchange, it arises from disorder, from noise: order from noise ..." [33].

Out of the tension of opposing forces (order and disorder, structure, and chaos), the system moves toward a few preferred global configurations out of infinite possibilities. The resistance to the post-treatment relapse of defects correlates to the range of variations the system can withstand given organisation and structure. Only living systems can integrate stresses, perturbations, and "noise," building ordered structures and following principles that lead to qualitatively different results. Where a form exists, we move away from homogeneity, disorder, and entropy. Any organic form constitutes an informational pattern about a spatiotemporal structure ready to pass to other informational, receptive structures. The decrease of biological complexity and the reduction of diversification results in the system's vulnerability and lower flexibility to changes in the environment. Many functions are achievable in multiple ways (redundancy) by activating parallel functional circuits. Thus, a treatment approach that determines the reduction of system characteristics may not be a good solution. A mechanical force operating on a tooth element for a time longer than six hours daily shift settles it into a new equilibrium condition. In addition, suitable stabilisation mechanisms must work to ensure stability at the new position since the forcing of the previous equilibrium has most likely pushed the dental system into a field of muscular or viscoelastic forces tending to return it to its starting condition.

Overspecialisation in facial features reduces resilience by eliminating backup functions and alternative pathways. It weakens decentralised functions, breaks down essential connections, and fragments vital components. On the other hand, morphological dynamics, characterised by opposing forces, help maintain structural balance. Organic structures comprise functional parts and repeating patterns of elements that contrast, cooperate, or compete. The principle of generation relies on discontinuity, where opposing natural forces stabilise matter. Empedocles suggested that life's foundation lies in the rhythmic interplay of attraction and repulsion, where cohesive forces counteract dissipation forces, anticipating the concept of material tensegrity. As René Thom notes [34], in any homeostatic process of an organic system subjected to stimuli in the long run, one of the corrective successions eventually imposes itself.

Nevertheless, in a situation of equilibrium between homogeneous elements, there is competition between imbalance-reinforcing polarities and balancing elements. Computational orthodontics can help visualise the hidden side of things and the diversity of constituent orthodontic processes.

In the same way, the realism of Hinduist philosophy recognises the multitude of every phenomenon of reality, breaking it down into six distinctive characteristics: substance, quality, action, generality, particularity, and adherence. This cognitive disposition resembles what the orthodontist must put in place in the diagnostic and prognostic process: it is of fundamental importance to find out everything involved to define as much as possible what is present but not yet expressed in the patient's reality. Computational procedures do not allow the operator to achieve transcendent states of knowledge driven by some form of mathematical theocracy. They help appreciate the multidimensional essence of each orthodontic process, unfolding in each orthodontic patient in a specific form. Predictive qualities describe the surplus of meaning possessed by each numerical parameter, which helps discover what lies behind morphological, nosological, metrical, taxonomic, relational, and developmental aspects.

Incorrect predictions, often labelled as "regression errors," stem from favourable and unfavourable factors already influencing but haven't fully materialised yet. In orthodontics, like in other medical fields, a practitioner's approach to new cases is shaped by existing knowledge. In unstable situations, each piece of knowledge loses relevance over time, decaying rapidly. We gradually depart from the original orthodontic data by employing repetitive computational procedures, where each cycle builds upon the previous ones.

References

1. P. Watzlawick, J.B. Bavelas, D.D. Jackson, *Pragmatics of Human Communication: A Study of Interactional Patterns, Pathologies, and Paradoxes* (W.W. Norton & Co., 2011)
2. D. Gentner, Analogical reasoning, psychology of, in *Encyclopedia of Cognitive Science* (Wiley, 2006). https://doi.org/10.1002/0470018860.s00473
3. J.D. English, T. Peltomäki, K. Pham-Litschel, Mosby's orthodontic review (Mosby Elsevier, 2009). Accessed 14 Dec 2023 [Online]. Available: https://books.google.it/books/about/Mosby_s_Orthodontic_Review.html?id=iabpHwAACAAJ&redir_esc=y
4. L.S. Horowitz, E.H. Hixon, *Nature of Orthodontic Diagnosis*, 1966th edn. (C.V. Mosby Company, 1966)
5. R.E. Moyers, F.L. Bookstein, K.E. Guire, The concept of pattern in craniofacial growth. Am. J. Orthod.Orthod. **76**(2), 136–148 (1979). https://doi.org/10.1016/0002-9416(79)90115-5
6. J. Mew, The dilemma of early treatment. Am. J. Orthod. Dentofacial Orthop. **133**(6), 784–785; discussion 785–6, Jun 2008. https://doi.org/10.1016/j.ajodo.2008.04.005
7. M. Blastland, *The Hidden Half: The Unseen Forces that Influence Everything* (Atlantic Books, 2020)
8. E.C. Olson, R.L. Miller, *Morphological Integration* (University of Chicago Press, 1999). Accessed 07 Oct 2023 [Online]. Available: https://press.uchicago.edu/ucp/books/book/chicago/M/bo3620375.html
9. R.S. Corruccini, R.M. Beecher, Occlusal variation related to soft diet in a nonhuman primate. Science **218**(4567), 74–76 (1982). https://doi.org/10.1126/science.7123221

10. N.N. Taleb, *Fooled by Randomness: The Hidden Role of Chance in Life and in the Markets* (Penguin, 2007)
11. D. Rasskin-Gutman, B. Esteve-Altava, Connecting the dots: anatomical network analysis in morphological EvoDevo. Biol. Theory **9**(178–193) (2014). https://doi.org/10.1007/s13752-014-0175-x.
12. J.C. Ramo, *The Age of the Unthinkable: Why the New World Disorder Constantly Surprises Us and What to Do About It* (Little, Brown and Company, 2009). Accessed 15 Dec 2023 [Online]. Available: https://books.google.it/books/about/The_Age_of_the_Unthinkable.html?id=tP-fPQAACAAJ&redir_esc=y
13. F. Jullien, *Traité de l'efficacité* (Paris: Grasset, 1996). Accessed 15 Dec 2023 [Online]. Available: https://www.persee.fr/doc/etchi_0755-5857_1997_num_16_1_1257_t1_0155_0000_2
14. S. Galella, D. Chow, E. Jones, D. Enlow, A. Masters, Guiding atypical facial growth back to normal. Part 1: Understanding facial growth. Int. J. Orthod. Milwaukee **22**(4), 47–54 (2011). Accessed 15 Dec 2023. [Online]. Available: http://www.ncbi.nlm.nih.gov/pubmed/22360082
15. T.F. Mulligan, Common sense mechanics. J. Clin. Orthod. **14**(7), 481–488 contd, Jul. 1980, Accessed 15 Dec 2023 [Online]. Available: http://www.ncbi.nlm.nih.gov/pubmed/6935237
16. C. Hertzog, J.R. Nesselroade, Beyond autoregressive models: some implications of the trait-state distinction for the structural modeling of developmental change. Child Dev. **58**(1), 93 (1987). https://doi.org/10.2307/1130294
17. H.F. Gollob, C.S. Reichardt, Taking account of time lags in causal models. Child Dev. **58**(1), 80 (1987). https://doi.org/10.2307/1130293
18. N.R. Sanders, L.P. Ritzman, *Judgmental Adjustment of Statistical Forecasts*. (Springer, Boston, MA, 2001), pp. 405–416. https://doi.org/10.1007/978-0-306-47630-3_18
19. R. Seising, From vagueness in medical thought to the foundations of fuzzy reasoning in medical diagnosis. Artif. Intell. Med.. Intell. Med. **38**(3), 237–256 (2006). https://doi.org/10.1016/J.ARTMED.2006.06.004
20. P. Auconi, M. Scazzocchio, E. Defraia, J.A. McNamara, L. Franchi, Forecasting craniofacial growth in individuals with class III malocclusion by computational modelling. Eur. J. Orthod.Orthod. **36**(2), 207–216 (2014). https://doi.org/10.1093/ejo/cjt036
21. R.S. Nanda, The contributions of craniofacial growth to clinical orthodontics. Am. J. Orthod. Dentofacial Orthop. **117**(5), 553–555 (2000). https://doi.org/10.1016/s0889-5406(00)70197-1
22. P.H. Buschang, J. Martins, Childhood and adolescent changes of skeletal relationships. Angle Orthod. **68**(3), 199–206; discussion 207–8 (1998). https://doi.org/10.1043/0003-3219(1998)068<0199:CAACOS>2.3.CO;2
23. B.A. Chvatal, R.G. Behrents, R.F. Ceen, P.H. Buschang, Development and testing of multilevel models for longitudinal craniofacial growth prediction. Am. J. Orthod. Dentofac. Orthop. **128**(1), 45–56 (2005). https://doi.org/10.1016/J.AJODO.2004.03.035
24. E.S.J. Abu Alhaija, A. Richardson, Growth prediction in Class III patients using cluster and discriminant function analysis. Eur. J. Orthod. **25**(6), 599–608 (2003). https://doi.org/10.1093/ejo/25.6.599.
25. H. Kitano, Systems biology: a brief overview. Science **295**(5560), 1662–1664 (2002). https://doi.org/10.1126/science.1069492
26. A.J. Seely, N.V. Christou, Multiple organ dysfunction syndrome: exploring the paradigm of complex nonlinear systems. Crit. Care Med. **28**(7), 2193–2200 (2000). https://doi.org/10.1097/00003246-200007000-00003
27. F. Jullien, *Nourrir sa vie* (Editions Seuil, 2005). Accessed 15 Dec 2023 [Online]. Available: https://www.seuil.com/ouvrage/nourrir-sa-vie-francois-jullien/9782020792172
28. A. Barton, F. Toyoshima, L. Vieu, P. Fabry, J.-F. Ethier, The Mereological Structure of Informational Entities (2020). https://doi.org/10.3233/FAIA200672
29. D.H. Enlow, T. Kuroda, A.B. Lewis, Intrinsic craniofacial compensations. Angle Orthod. **41**(4), 271–285 (1971). https://doi.org/10.1043/0003-3219(1971)041%3c0271:ICC%3e2.0.CO;2
30. C.H. Tweed, A philosophy of orthodontic treatment. Am. J. Orthod. Oral Surg. **31**(2), 74–103 (1945). https://doi.org/10.1016/S0096-6347(45)90163-3

31. K. Kelly, *Out of Control: The New Biology of Machines, Social Systems, and the Economic World* (Addison-Wesley, 1995). Accessed Dec 12, 2023 [Online]. Available: https://books.google.it/books/about/Out_of_Control.html?id=mGZfZKtsxtwC&redir_esc=y
32. D.J. Hand, *Dark Data: Why What You Don't Know Matters* (Princeton University Press, 2020). Accessed 15 Dec 2023 [Online]. Available: https://www.amazon.com/Dark-Data-What-Dont-Matters/dp/069118237X
33. A.F. De Toni, L. Comello, L. Ioan, Auto-organizzazioni: il mistero dell'emergenza nei sistemi fisici, biologici e sociali (Marsilio, 2011)
34. R. Thom, *Structural Stability and Morphogenesis* (Perseus, 1979). Accessed 15 Dec 2023 [Online]. Available: https://www.scribd.com/document/490973990/Thom-Rene-Structural-stability-and-morphogenesis-an-outline-of-a-general-theory-of-models-Perseus-Books-1989-pdf

Chapter 6
Orthodontic Digital Phenotype: Make Data Mean More

Artificial Neural Networks allow outcomes to be related anyway to the input features by multiple layers of logistic regressions and iterate adjustments of feature weights to minimise the error between predictions and actual labels ("backpropagation of error"). The multi-layered arrangement of Artificial Neural Networks is a filter that works from gross to subtle, increasing the likelihood of detecting a correct diagnostic or prognostic response, capable of automatically learning an implicit representation of patient data.

During the learning process, the weights between artificial neurons and the strengths of connections change. When the weight between two connections is small, the network lets the signals flowing in that path fade and fade out because the path is unlikely to contribute significantly to the final solution. Otherwise, the connections between neurons tend to strengthen and transmit values to the next layer as a summation of the values present in the previous layer. Inputs are creatively mixed within the network using weights and activation functions. We can establish a mapping function $y = f(x)$ from x (the data we possess) to y (the data we want to predict from x).

A reliable mathematical model requires the identification of a few essential variables under the assumption that the neglected aspects will have a modest effect on the relevant characteristics and the system's dynamics. At the outset, the critical moment in the modelling process is transitioning from observing the specific problem of interest to identifying the most representative state variables. Computer *simulation* allows us to focus on particular characteristics and to predict their dependence on additional parameters, having chosen several variables capable of constituting a model. The next step is formulating hypotheses for the system's behaviour under stress.

In a nutshell, we can express a malocclusion picture in the form of:

(i) Teeth-bone bases size disproportion.
(ii) Uncoordinated development of the bone bases.

P. Auconi et al., *Systems Orthodontics*, Understanding Complex Systems,
https://doi.org/10.1007/978-3-031-64932-5_6

While a system is a relation of characteristics, a complex system can be considered a relation of systems in which the overall system exhibits behaviour distinct from the parts. The crucial point is the definition of the interactive schema and its ability to indicate an architecture. The macrostate (the phenotype) reflects microstate situations (expressed as a network interaction of local features); however, the same macrostate may correspond to different, indistinguishable microstates.

It is, therefore, necessary to search for other qualifying features, (i) to visualise and locate possible highly connected local working modules, (ii) to determine the extent to which we respect the autonomy domain of each module, (iii) which module has taken over, whether the balance between the subsystems of the growing structure remains constant, (iv) to locate the information condensation zones. By observing the dynamic totality of a phenomenon, the computational approach visualises its systems constitutive matrix, leading to a diagnostic schema. Logical-mathematical schemes help assemble clinical and radiological data in a progressive diagnostic enrichment necessary to support or disprove initial assumptions, keeping in mind that each diagnostic categorisation involves decreasing the system's complexity. For the algorithm to orientate correctly, the training examples and predicates given by the domain knowledge, i.e., the background, are helpful. Liu and co-workers developed a technique that eliminates uninteresting correlations between parameters and identifies the most frequently followed association rules, indicating the phenomena' regularities. We can represent the essential dynamic structure of the system in this way (we learned about this construct when discussing *propensity*). The secret is to prune global rules that are too specific or capture idiosyncrasies in the data to arrive at meaningful subspaces [1]. The technique is similar to the pruning of Classification Tree bifurcations to avoid overfitting. In any case, one must compare the computational findings to what an expert expects.

Studying the number of connections each node has in the feature network helps clarify essential aspects such as robustness, structural balance, and resistance to recurring disharmony after treatment. Malocclusions represent variations of disharmony influenced by genetic, familial, environmental, and other unknown factors. Genetic influences primarily affect early skeletal malocclusion components, while local morpho-functional balance factors regulate tooth position-related issues. Modern malocclusion interpretation considers diagnostic assessments across all spatial dimensions. Discrepancies in sagittal and vertical dimensions of teeth and skeleton influence each other. Given the diversity among patients, orthodontists must develop versatile concepts. Each data point contains its antecedent, consequent, and connection between them, and clinicians decide the observation level, which also implies assumptions about growth quality over time. Mapping dentoskeletal structures should focus on relational patterns rather than isolated facts. Clinicians create models to understand relationships in clinical events, considering initial conditions, patient-specific occurrences, and other factors impacting the outcome. Nature, creative by necessity, has developed various adaptive structural solutions that have allowed the evolution of animals, plants, and viruses for four billion years. Conscious emulation of nature's best strategies reveals an enormous variety of inspirations. Orthodontics is also the realm in which order, disorder, structure, and variety exist during facial growth. The challenge present in several research fields is to find and follow the "laws of disorder" to identify hidden connections underlying the dynamics of phenomena that are haphazard, i.e., the unexpected mutation of facial features during growth.

Most up-to-date predictive machines are so effective at considering complex interactions between different indicators that they can even perform "counterfactual" simulations or visualise the effects of alternative growths and treatments. In a similar way to those used by economists in financial forecasting: "What would have happened if things had gone differently?", "what would have happened if I had acted differently?"

6.1 The Problematic Interface Between Machine Learning and Orthodontics

In the past decade, we collected considerable data to develop predictive electronic systems in clinical and research orthodontics. The promise was like what was hypothesised in medicine: to optimise patient care, you can exploit the memory of the best clinical decisions made by thousands of professionals (Table 6.1).

When dealing solely with numbers, there's a risk of losing the human touch, but humans possess a unique advantage: common sense. Humans can understand the reasons behind processes and perceive background knowledge, abstract similarities, and analogies between patients based on limited cases. In contrast, Machine Learning needs this inherent understanding and relies on data-driven correlations rather than theory. Computational tools can predict future characteristics when they represent the underlying data structure. Researchers and orthodontists can analyse data domains using these models. Anomalies cluster around major skeletal defects, reflecting non functional disharmonies. Feedback between adaptation sensors and growth actuators provides morphological information. Computational machines don't require a deep understanding of patient phenomena; successful models rely on vast data. They seek regularities within data, prioritising numerical patterns over clinical hypotheses. However, machines should also grasp qualitative aspects of data, recognising the complexity of facial structures, including various components and growth directions. Time influences disharmony differently pre- and post-puberty, emphasising

Table 6.1 Translating from daily orthodontics to machine learning models, the orthodontist might

1.	Identify hidden craniofacial trends in large datasets of growing patients
2.	Leverage trends to make growth outcome predictions
3.	Compute the probability for each possible growth and treatment outcome
4.	Clarify the effects of the co-occurrence of skeletal defects and renormalisation phenomena on growth and treatment outcomes

general factors earlier and local factors later. Orthodontic information is inherently complex and imperfect, reflecting the unpredictability of skeletal development. Developmental noise contributes to phenotypic diversity, even with fixed genotypes and environment.

Nevertheless, following Chris Lanton's suggestion, "...complexity science teaches us that the complexity we see in the world is the result of hidden simplicity..." a complete description of phenotype-genotype mapping must separate the stochastic character of development from the general growth rules. That is, from what we should expect (with some degree of probability) (Table 6.2). In the craniofacial system, the macro-scale morphological phenomena emerge from the aggregate behaviour of the local interactions of the micro-scale components.

In medicine, diagnostic errors are related to too hasty closure of the diagnostic process, a tendency not to consider alternative hypotheses to the first diagnosis ("anchoring bias"), the tendency to consider diagnoses that are easy to remember ("availability bias"), the tendency to think only confirmatory data to initial diagnosis

Table 6.2 Network metrics are the more elementary quantities in network theory that allow the characterisation of the orofacial system to start

Network quantity	Definition	Meaning	Example
Vertex	The essential element of a network denoting one of the features of the orthodontic system	The object under investigation,	A distance between Co and Go points, or an angle. Any orthodontics feature
Edge	The relation between two elements of the network	The rule that gives the interaction between the vertices	A correlated variation concerning the expected values in a patient
Distance	The minimum number of edges to travel to reach a vertex	If edges mean correlation, how far apart two vertices are	
Degree	The sum of the edges entering a vertex	The importance of this vertex in the system	
Clustering	The number of neighbours that are connected		
Closeness	The average distance to the other vertices	How far is the vertex concerning the other vertices in the system	
Betweenness	The number of times	How crucial is the vertex to keep the system connected	
Community	A subgraph is more densely connected inside than outside		

while ignoring contradictory data ("confirmatory bias") [2]. Conversely, the error rate for machines mainly results during the learning step. The most common cause lies in poor quality training data, such as irrelevant features, spurious associations, false assumptions, inappropriate patient attributions, and indications that cannot represent the patient's clinical background. Applying the same statistical rule to all cases, the prediction made by the algorithm tends to minimise the noise. When clinical and cephalometric data on a patient's history are incomplete or too varied and scattered, an algorithm that leverages and integrates diverse information can still describe them. But, of course, the answers can only be probabilistic.

Machines tend to complete information about reality when only part of it is known. Rigid models define a reality of their own (the model) and use it to justify their results and to make predictions. Statistics rely on large numbers of data to compensate for exceptions and anomalies. With stable conditions and continuous feedback, statisticians can refine their models to a certain point. Unequal probabilities due to morphological constraints, competition/collaboration strategies, emergent phenomena, and more contribute to unequal developmental possibilities. Thus, living matter is not reducible to a mathematical-statistical formalism suitable for an elementary mechanism. The more the presence of a particular sign in a message is, a priori, improbable, the more its occurrence is, a posteriori, informative. Some developmental, morphological properties directionally constrain evolution paths because they define the limits of the possible variations associated with that specific morphology. While it is true that everything relates to everything in the dentofacial system, some subspaces are in a much closer relationship with each other than others. Bayesian networks have clarified not only the correlation relationships between the various factors in the system but also the direction of influence that one parameter exerts on another. Influence is not necessarily exercised through relationships of spatial contiguity but also at a distance. Moreover, chains of cause and effect determine that a stimulus's impact can be determined long afterwards (see Figure page 76).

The Bayesian network approach provides a way to perform a preliminary verification of the hypotheses targeted by those trials to allocate resources efficiently [3]. Finally, they incorporate both prior knowledge about the variables of interest and the uncertainty present in the data. To visualise the influence of factors non-directly quantifiable through clinical or radiological findings, it is possible to subject the data to algorithms capable of visualising the latent variables underlying the variability of the values observed in the dataset. In other words, it is possible to define an errors-in-variables model, i.e., a regression model that considers the biases induced by measurement errors in the independent variables. Linear regression models assume that it is possible to measure the independent variables of the system without making errors.

Furthermore, the error on this measure depends on the estimate of the dependent variable or response. When variables cannot be estimated accurately (as in the case of some cephalometric landmarks), regression coefficients are inaccurate and usually underestimated ("attenuation bias"). The multidimensionality of dentoskeletal characteristics makes it necessary to consider the influence of inaccuracies in the final response of the algorithm (dependence on disparate difficult-to-perceive factors of functional nature, habits, growth, complex correlations between craniofacial areas, and so on) [4]

Algorithms learn from past data, but finding connections between features only sometimes leads to accurate future predictions, especially when dealing with morphological and functional interactions, not just statistics. The average traits of similar orthodontic patients may not apply due to individual skeletal variations and changing environmental dynamics. Information persists in physical structures like bones and joints. Orthodontic patients carry hidden data and variables not expressed numerically; about half of an organism's characteristics may be noticed. Human learning involves qualitative interpretation, while machine learning relies on probability without prior knowledge. Orthodontic features evolve, affecting long-term model stability. Recognising subtle changes is vital for accurate representation. Feature engineering transforms raw patient data into numeric formats, enhancing model accuracy by integrating quantitative and qualitative aspects. It involves understanding patient features, incorporating operator knowledge, and considering computation pros and cons. Cephalometric variables are often better expressed as deviations from population norms adjusted for age and gender.

The programmer must acknowledge uncertainty and bias in the data to prevent overly idealised models. Excessive cleaning of orthodontic data simplifies reality. The learning machine needs diverse examples of subjects with different skeletal issues and standard features across ages to predict malocclusion early. It also requires a series of features from subjects with varied growth patterns to ease modelling challenges. Considering how features occur together is crucial, but the machine can only grasp some possible skeletal and dental realignment phenomena. Realistic models demand collaboration between clinicians and computer scientists. Challenges in malocclusion mapping to Machine Learning include mathematically defining overall skeletal imbalance, understanding each characteristic's biological essence, weighting expressive patient features, choosing analytical methods for growth, habits, and therapy response, and capturing latent dimensions of unsuccessful treatment. Facial modelling for dentoskeletal evolution prediction poses further challenges. Understanding adaptation to dysmorphosis and distinguishing it from other biological processes affecting growth trajectory is essential. We often need an understanding of causal pathways, making assigning weights to these phenomena challenging. Acknowledging the system's spatial structure is crucial despite limitations in fully understanding connections and influences among features.

The chairside orthodontist uses a Bayesian approach to update their understanding with new patient data, gradually refining their probabilistic decisions. Humans struggle to make probabilistic decisions with partial information and don't calculate data interactions precisely. Feature engineering helps identify specific characteristics impacting craniofacial growth by assigning weights to each feature's influence. In patient datasets, specific components like dentoalveolar remodelling may need to be fully recognised by Machine Learning. When the algorithm's model diverges significantly from clinical logic, it challenges human-computer interaction. Orthodontists must guide prognostic and treatment decisions, as machines lack clinical backgrounds and struggle to apply learned knowledge to new situations. While machines rely on numerical empiricism, orthodontists possess native clinical experience. Simple clinical situations usually align well between human and artificial interpretative models, but complexities arise in progressive contexts where knowledge decays over time. Human memory involves encoding, storing, and retrieving information. In orthodontics, humans manage the noise in the information based on the patient's context and the operator's previous knowledge. We simplify the world by categorising it, creating intuitive clusters of features and associating probabilities with outcomes in each cluster.

Researchers and practitioners must understand the basis of algorithmic feature selection, decisions, and predictions to disseminate best practices and trust in computational procedures effectively. This action requires grasping the differences between numerical and orthodontic formalisms. Machine Learning optimisation relies on mathematical principles like mean squared sums, regularisation, normalisation, and gradient descent. In contrast, dentoskeletal optimisation primarily involves adaptation.

6.2 Personalised diagnosis

In biomedical fields, diagnostic devices sometimes match the human brain's ability to handle information. Orthodontists need help with the discrepancy between the brain's capacity (limited to seven variables) and the complexity of orthodontic patient data. Predicting outcomes involves estimating probabilities and testing hypotheses. Neural Network procedures apply hidden layers between input and output, which help but are not easily interpretable. Understanding why certain patient features relate to specific conditions remains difficult. Despite efforts to understand interactions at a micro-level, establishing similar groups of orthodontic patients is challenging due to subtle differences in growth and development. Probability theory addresses uncertainty in orthodontics, but data mining algorithms can magnify uncertainty. Different models and biases further complicate diagnosis, especially with patients falling between diagnostic categories. Measurement errors and biases in data distort decision-making, with algorithms striving for the best option amidst uncertainty. As complexity grows, reliance on numerical data increases, overshadowing empirical and clinical aspects. Theoretical and practical gaps lead to noisy data, while hidden algorithm nodes struggle to interpret transitions from phenomena to numbers.

Our instinct to control leads us to create simplified models of complex phenomena, standardise characteristics, and eliminate unnecessary data. We aim to uncover individual growth potential present at the start of development. However, predicting how initial characteristics will evolve over time and whether models will accurately represent phenomena remains challenging. Error-based learning in models can lead to significantly wrong outcomes. We rely on probabilistic views of reality, but the unpredictable nature of skeletal growth limits stability. The lack of repeated examples and similar developmental trajectories makes generalising from patient groups to individuals difficult. We can't fully interpret every aspect of living matter or manipulate all causes and effects. Forecasting errors in growth and treatment response are possible due to the limited inclusion of past and future models focusing on the present. Crossing technological boundaries in prognostic processes may lead to new orthodontic theories powered by technology. While machines lack intuition, computational models enhance professional skills in orthodontics and will continue to do so in the future.

6.3 Networks Orthodontics

Since both the global structure and the nodes have an effect of mutual influence and mutual structural constraint, once we establish the worldwide propensity, it is easy to define the unfolding of the system probabilistically. It is necessary to clarify the virtuality of the system, as well as the initial conditions, to obtain this information. The orthodontist must strive to perceive the continuity of the transformation, i.e., the transition process. Systems thinking disregards the changing 'things'; it concerns the inapparent transition of the whole system, with the change "in which each moment is contained in the next" [5]. In the fluidity of growth, the initial phase of a transformation already contains its unfolding. The network topology helps define the covariances in dentoskeletal organogenesis: a balanced brachiocephalic system may have specific co-occurrences that, sooner or later, expose the subsequent variables towards the onset of disharmony. Or latent factors that we need to unearth. It is necessary to highlight all the virtualities contained in the initial condition to define the certainty of the unfolding of the process. Determining the pattern at time T-zero means restricting the probability of the pattern at T-one to the compatibility of future trajectories with the present structure because the potential for future development is already present in the early data. In the transition, all internal separation vanishes. When does T-zero end? When does T-one begin?

We shall consider the situation and developments separately. We need help seeing things simultaneously in transformation and their configuration [5]. In living systems, things are implied by what they are: events and structures are not separable. As stated, we must strive to abandon the concept of causality ('things take their truth from causes'), hence 'explicitness', and enter the implications of the system into the propensity. In statistical regression, we distinguish between the regressor (the explanatory variable contributing to varying degrees to the dependent variable) and the regressed. In the correlated logic of events, the propensity of phenomena teaches that the regressor is also regressed and that the explication is an *implication*. Events evolve by their internal configuration. Future variation is already contained in the present morphology, as it is just a matter of discovering it [5]. The configuration is already inclined to take a direction by a transformation becoming more and more probable. In preparing the treatment plan, strategic systems thinking turns to evaluating the situation's potential and the concomitant favourable and unfavourable factors. It follows the branching of thought to determine the importance of the variables as a function of advantage. The more therapeutic transformations are present in the situation's development, the less they are noticed. One must avoid being driven by impatience or inertia when deciding when to intervene. Letting the natural generation of the process run its course and letting an effect mature without interfering does not mean neglecting it. To clarify the patient's underlying situation, one must conform to the propensity and support its unfolding.

Once we have gained a deep understanding of the overall structure of a network, it's easy for us to investigate its dynamics. What we want to know is not a static roadmap. We must understand how components interact to form the structure, how a change to one local component may affect other parts, and how the local components interact dynamically. Given its importance in defining the topology of orthodontic processes, it is necessary to return to the subject of Network analysis. A general way to represent complex biological systems is to describe them using the most straightforward architectural units. A network, or in more formal language, a *graph*, is a simplified representation that reduces a system to an abstract structure capturing the basis of the system's connection pattern [6]. The network can identify which problems and constraints pass in the skeletal system and the sectors in which constraints and self-repair dynamics will be active. It locates empathetic relationships between facial components, the resonance of the system's vital processes (eruptive conflicts, muscle tone modulations, structural imbalances, structure translations, and so on), and how these influences are transmitted. Furthermore, network analysis allows us to visualise and map various systems where other metrics are unavailable.

Networks (Fig. 6.1) are a way to represent complexity. As stated, the most straightforward possible network representation reduces the system's elements to *nodes* or *vertices* and their pairwise relationships to *links or edges* connecting pairs of nodes.

In so doing, the relationship between the parts of the system can be efficiently quantified. In addition to the points above, Networks help us define:

- The **degree** is the number of edges per vertex. The degree is the most straightforward topological index. It allows an immediate evaluation of the regulatory relevance of the node. In medicine, an edge is often a correlation between two distinct features (the nature of this interaction determines the significance of the structure). In the figure above, the degree of node i is 1.

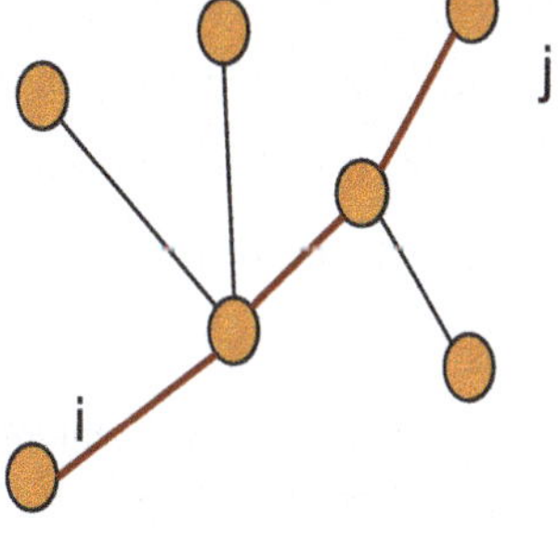

Fig. 6.1. A simple network

- A **distance d(i, j)** is the minimum number of edges one travels from vertex i to vertex j. The simple example in the figure is the red path composed of three edges.
- Once we have a distance, we can define how far every vertex is distant from the others. The lesser the average distance, the larger the **centrality** of the vertex. Various centrality measures are possible: the closeness (the sum of the inverse of distances with all other vertices) or the betweenness (how often we pass through an edge/vertex when going from one vertex to another one. The betweenness in biological terms can indicate the relevance of a component as functionally capable of holding together communicating components.
- A **cluster, community, or module** of vertices that is a structure denser (more edges) than the average density of edges of the whole graph. Again, in the case of medicine networks, this refers to vertices (features) strongly correlated with each other.

Let's see some examples related to orthodontics. The growing craniofacial system is an aggregate skeletal structure of various agents. The clinical characteristics are the network nodes, and the relations between these characteristics are the links among such nodes. The nation-mention length is the bridge that connects skeletal segment data with angle data (see Fig. 6.1).

Edges or links can represent functional, anatomical relationships between the nodes, such as "catalyse", "bind to", or "shift"). The network's inference and analysis refer to information on a system's identity and state of elements to their functional relationships and the extraction of biological insights and predictions. When a network is highly connected internally, the behaviour of the whole governs the operation of individual variables. At first glance, in Fig. 6.2, some nodes appear to have more links than others. A more accurate analysis reveals that sometimes a critical role can be played by nodes, which are crucial intermediary nodes; most of the information circulating in the system must flow through these nodes. If such a node disappears, the network splits into smaller groups of nodes that no longer communicate. This property of centrality makes this quantity particularly important to determine the "importance" of a vertex in the structure. Central nodes play a control role (intermediation, or "betweenness") among other nodes in the network (Table 6.1)

The various quantities describing the network are related to the system size needed to explore and compute them.

1. *Local Parameters*
   a. The degree (the number of edges per vertex)
2. *Nearest Neighbours Parameters*
   a. Clustering (the number of triangles)
3. *Global Parameters*
   a. The density (total number of links divided by the maximum number of possible edges)

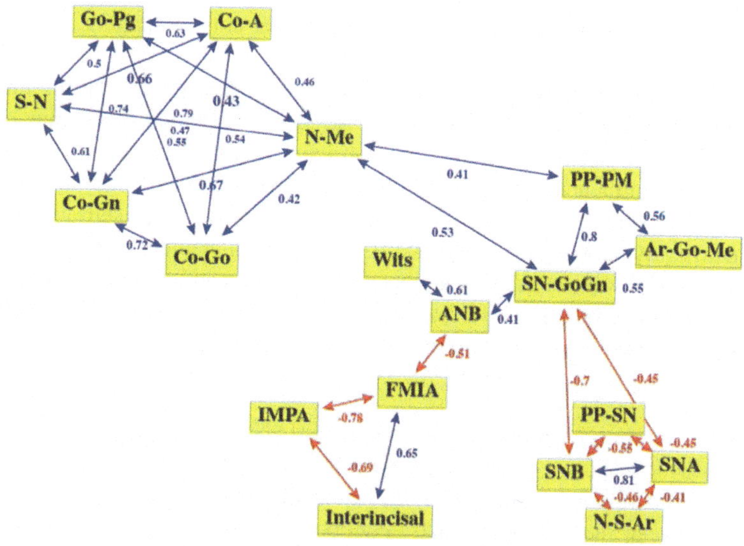

Fig. 6.2 An example of network definition starting from craniofacial features as vertices and correlation between these elements as edges. In addition to geometric-type relationships between nodes belonging to neighbouring structures, correlations also express functional interconnections. In this network, the ANB node, which marks the extent of sagittal imbalance, is interconnected with dentoalveolar renormalisation nodes IMPA, FMIA, and Interincisal. The interaction between features can generate harmful feedback mechanisms, which tend toward rebalancing, and positive feedback mechanisms, which cause the speed of interaction to increase and recursive relationships between components, with effects disproportionate to causes, which put the system in a critical condition of possible rapid imbalance during growth. (From Scala A., and coll.)

b. Closeness
c. Betweenness
d. Community.

Network structure and the pattern of interactions can have a crucial effect on the system's behaviour: the architecture of relationships matters, not the precise identities of the elements themselves. Network analysis allows us to characterise some system features under examination, i.e., robustness, distance between nodes, and number of links for each vertex [7]. The *degree* of a network corresponds to the number of nodes adjacent to a given node. A node with a very high degree interacts with several other nodes; it is likely to be a regulatory hub of the network. The *shortest path* between two nodes underlines how far apart, e.g., dissimilar, they are in the structure. The *mean shortest path* of a network gives the overall facility with which the nodes exert their reciprocal functions. *Bridge nodes* are nodes whose absence would split the network into two separate parts. They allow the detection of different subsystems and represent the connection among them. *Closeness centrality* measures the information diffusion from a given node to other nodes. It indicates the relevance of a node as functionally capable of holding together communicating nodes of the system.

Networks offer a formalised model to display a process by converging multiple factors rather than according to causal processes. In orthodontics, by locating mutual relationships among craniofacial features, structural imbalances, eruptive conflicts, and how these influences propagate. It can give meaning and identify which information and constraints are transmitted and the sectors in which self-adaptive and self-repairing dynamics will be active. It is a significant challenge for orthodontists to integrate these various datasets to generate diagnostic and prognostic hypotheses (knowledge-based hypothesis formation has been a focus of artificial intelligence: the search for hypotheses as "pattern in data") (see Box 5.1).

Box 6.1. Finding Meaning in the Orthodontic Network The growing craniofacial system is an aggregate structure of various agents whose clinical (radiographic, functional, etc.) characteristics are nodes connected by links (the relations between these nodes). Each network component acts as a processing unit of information. A network structure makes some orthodontic phenomena treatable mathematically in terms of how the agents are related rather than what they are composed of. A network must include all distinct entities that can participate in the relationship under study. Providing a powerful representation of the interactions and highlighting the global interdependence, Graph structure can have a significant effect on the behaviour of a system. It is the pattern that matters, the architecture of relationships, not the precise identities of the elements themselves.

A core set of organising principles characterises networks operating in many biological, technological, or social systems. Understanding malocclusions in the context of these network principles allows us to address some principles involved in developing craniofacial dysmorphosis. Figure 6.2 shows the interconnection network between craniometric radiographic characteristics of 250 female Class III patients (14–16 years of age). The network representation instantly visualises the relationship structure between variables (correlation/anticorrelation coefficients). But what good is that? First, you must answer these questions: what are the most critical nodes of the network? Some features interact more effectively. What are the essential nodes of the network? What is the path that allows relationships to be more significant and the exchange of information faster? The answer to these questions lies in the network structure and how the agents are connected. At first glance, the graph lacks a well-defined structure: there are no quantitative entities and no numbers other than the notation on the correlation coefficients. But, in hindsight, we see areas of thickening of nodes and zones of preferential connection with a more significant number of correlations between nodes. In these areas, the logical-mathematical analyses can indicate some underlying rules of interaction, relationships of clinical co-occurrence, cooperative dynamics, kinetic parameters, basins of attraction, mechanotransduction barriers, hypothetical growth vectors, and so on. In the example, NMe and ANB nodes exhibit high *betweenness* scores in the network, i.e., crucial intermediary correlations. We can calculate the intermediation capacity between system nodes, a feature that confirms their remarkable ability to govern the entire network. We also noted that the bottom right and top left nodes of Figure 6.2 have a dense interconnectedness: this could indicate the liability to exchange information quickly. The network decomposition into motifs is often helpful in understanding the functional organisation of the system (many biological complex networks display a hierarchy with several interlinked sub-graphs) [6].

In the modular organisation of a complex system, each node "reads" information from neighbours according to the propagation characteristics within each node and responds to stimuli, modifying its steady state. It is necessary to establish the location of the areas of thickening of specific balancing or unbalancing characteristics to appreciate the network's vulnerability level, such as the likelihood of incurring the variation of facial growth characteristics. The phase transition of a node of the network may depend on the activation level of the neighbouring nodes and the degree of connectivity with these nodes. A node associated with many other nodes usually exhibits high stability because every variation of an adjacent node has a negligible effect on the other nodes due to the dilution of the message. When the nodes are too weakly connected, the network tends to fragment into "islands" of nodes. In case of fragmentation of nodes, the networks provided a wide range of well-distributed, solid and balanced connections trapped in a "stable state".

A living system's evolution reflects the parameters imposed by the attractors and the effect of "emergent" collective variables displayed in processes where all nodes play a role. These global processes emerge through network metrics such as "average path length", "distance", "diameter", and "stress". The network algorithms capture some aspects of the "organised complexity" of the system, providing a probabilistic approach to the evolution of growth processes, clinically otherwise not knowable. They achieve that by crossing the identification of the local dynamics (under the effect of specific local attractors) with the global dynamics of development (main structures, patterns, regularities).

6.4 Search for Organising Principles

As the craniofacial system is a singularity that produces singularities, the details of their evolution are usually related to processes that we cannot perceive in their totality. What appears at the end of the facial growth process results from an ongoing, often ineffective, constant adaptation of the parts to changing circumstances. Hence, the diagnosis and evolutionary prediction have a significant degree of uncertainty. Some characteristics tend to affect others more than others. Some aspects tend to grow and acquire importance, while others stand aside and have a negligible role in the development of the system. The complexity theory tells us that in the tangle of connections, the more linked characteristics (in network analysis, the nodes with high "degree", "centrality", and "betweenness") have a more significant effect in determining the evolutionary behaviour of the whole system. It is the consequence of a specific network dynamic called "preferential attachment": in a growing network, incoming nodes prefer old ones with many links. The higher the degree of a node (and therefore the older it is), the higher the probability of attracting a new one. Most of the time, the most connected nodes ("hubs") will be much more attractive, and occasionally less connected nodes will receive new links. At the beginning of the growing process of a network, all the nodes have the same number of links. Subsequently, some nodes start to accumulate more and more links than others. In those above "rich-get-richer mechanism", the initial minor differences in connectivity are progressively amplified. Thus, the connection mechanism of the incoming nodes is not random, as the network becomes progressively inhomogeneous. Understanding how single and group variables are organised in the network—topologically and in terms of functional dependencies—can lead to a better knowledge of the coordination of developmental processes.

Orthodontic evolutionary processes are not always reducible to stable patterns. Biological processes have both a proximate and an evolutionary cause; each previous event contributes to subsequent events. A physical system lives between a past that accumulates and a future that develops by imposing itself into the present. It forms the structures that create the conditions for its further growth. Suppose the future is a consequence of the past. In that case, we have seen how many attempts there have been to refer to some general rules of complex nonlinear systems, such as the capability to store previous adaptations for future use. In the evolutionary probability inherent in the sequence of internal configurations, certain groups of configurations become strongly connected and stable (as we have seen, network analysis highlights some orofacial characteristics converging into attractors) and, for this reason, persist more likely than others. These "emergent" phenotypes are the effect of the "preferential attachment" phenomena during evolution. They define the pattern of balance that does not stimulate incentives to change during the growth process.

All complex systems tend to resolve the complexity into a series of stable interactions leaning towards some form of order. These processes depend both on the interaction of general mechanisms that operate as constraints (the "laws") and on the variety and space-time specificity of local events (the "singularity"). In a physical system, only current forces can lead to modifications of everyday phenomena. In biology, the past and the future may influence the current mode of development through "surrogates", representing them in the present circumstances.

Everything not explicitly excluded by the clinical and radiographic information available has a certain chance to occur. The transfer of the biological optimisation concepts into the orthodontic field does not look for one answer but for a probability distribution of different scenarios. We must start from the choice of the probability distribution that would be obtained by summing all the possibilities allowed by the auxologic laws of organogenesis. As we mentioned in the botanical framework (investment in height strategy improves access to light at the expense of neighbouring plants), the teeth's payoff, which translates into oral fitness, depends both on individual and overall teeth strategy as regards the space during the transitional period [8, 9].

The outcomes of these growth strategies are visible in orthodontic patients: each tooth carries some orofacial characteristics that are part of the fitness landscape of another tooth. The transfer to orthodontics of the concepts of complexity and systems thinking intends to avoid bringing further sophistication to a discipline whose procedures and skills are already complicated nor to suggest abusive analogies or pretentious shortcuts that ignore the complex details of orthodontic methodology. Just as happens in disparate fields, from planetary financial crises to global warming, religious tensions to pandemics, some problems seem to evolve naturally towards a state of unnerving "self-organised criticality", spreading a vaguely oppressive climate of restless change and global instability. Faced with the insolubility and vagueness of many complex problems, we must find a new way of thinking to handle this worrying interdependence.

Computers can only provide reliable answers if they deal with precise facts. An utterance must be either true or false; it must follow binary logic, and there must be no uncertainty. Fortunately, doctors have common sense (they possess an inherent fuzzy logic); they know that many painful symptoms are nuanced, and many truths are only partial but no less critical for diagnosis and treatment plans [10].

The orthodontist must bracket nuanced assumptions about the clinical picture, such as "singularity", "locality", or "wholeness" of the symptoms, "temporality" or "permanence", "co-presence" of disparate signs, not always related or indicative of a particular facial dysmorphic type to achieve diagnostic synthesis simple enough to allow the rapid development of a treatment plan. Sometimes, situations related to the superimposition of a static diagnostic scheme on the biological dynamic diversity of the facial system subtends phenomena of clinical misinterpretation, over-diagnosis, and overtreatment.

Since things are implied by what they are, evolution is at one with morphological constitution. In this form, the orthodontist must recognise the process and internal logic of developing morphological phenomena. He must see things simultaneously in their configuration and transformation. The prognostic error between prediction and reality, between theory and practice, is not just a noise, a statistical discrepancy; it is the conceptual mismatch between what is a morphological snapshot and what pertains to the narrative of the situation. In orthodontics, what is most stable survives and grows stronger over time; what is unstable tends to disappear. But to what extent can our mental modelling perceive the accumulation of past accidental events, the selective advantage of the division of labour between parts, and the countless latent variables in the system?

The stomatognathic apparatus has multiple sensors detecting structural and functional disharmony, capable of triggering a wide range of rebalancing processes. The difference in sagittal growth between maxillary and mandibular influences the determinism of the correct molar relationship, beginning in early mixed dentition. The period of mixed dentition is a critical period for compensations. Parallel to the significant interindividual variability in growth patterns, we should expect great variability in natural compensatory effects and in the therapeutic effects we can achieve by exploiting them. The therapeutic "unblocking" of a Class II division 2 malocclusion offers an example of the cause-and-effect phenomenon in the course of treatment: the linear sequence of increase in transverse diameters of the upper jaw, anterior repositioning of the mandible, improvement in breathing, swallowing, phonation, lingual and antigravity postural pattern. But, as always, the nonlinearity present in the phenomena of complex systems teaches that virtuous release of natural compensatory adjustments occurs mainly in subjects with a whole propensity of favourable auxological pattern and normal muscle tone. Examination of the cephalometric tracing reveals the subtle functional synergies put in place by the stomatognathic apparatus to intercept best and sort occlusal loads: the parallelisation of the principal axes of the molars and premolars and the connection of these with the lines of force of the jaw-elevating muscles (Fig. 6.3).

These mechanisms of natural ergonomics ensure maximum functional benefit, minimum energy engagement, least detrimental effects on dental-periodontal structures, joints and muscles, shunting of occlusal load to adjacent elements, the neutralising the excessive occlusal loads, and distributing on opposite cuspal slopes of point contacts. Despite these self-organising mechanisms, a slow but steady mesial drift of the arches has yet to be prevented, making progressive crowding inevitable.

As is well known, the ideal occlusal function must be able to rely on a convergence of features in which no one feature overrides the others. It must also ensure a certain degree of functional freedom, compensation, and mutual harmony of curved surfaces

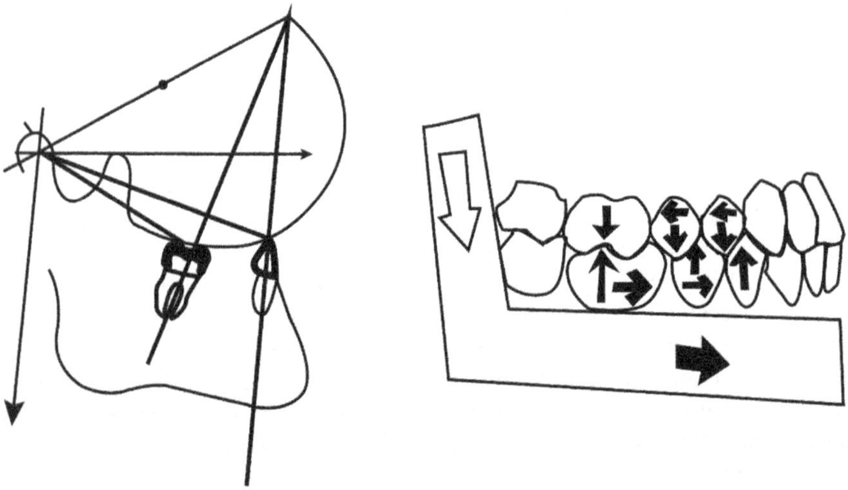

Fig. 6.3 Each tooth element tends to assume the most geometrically convenient position to intercept and dispose of the forces felt. The central axis absorbs the vertical pressures with minimal engagement of the paradentium. The direction of the pressures on the individual elements is steeper at the incisor level and more inclined mesially at the molar level. The fulcrum of the system is the condyle. Over the years, the molars, premolars and canines tend to compact their principal axes to optimise group function, ensuring tight interproximal contact points (Modified, from: [11])

arranged in various planes. Even in markedly crowded situations, these curved interocclusal surfaces ensure maximum matching of shear and drainage zones. One more lesson from nature: quality is present in the variability of global functions and local geometric configurations, even in seemingly irregular and optimal connections.

6.5 Action Does Not Always Equal Reaction

The biomechanical laws that govern orthodontic dynamics are not laws of action-reaction but of stimulus-response or action-adaptation. Reaction to stimuli is a response that induces structural changes and local or global (cellular, tissue, organ) responses to conquer new patterns of organisation. During treatment, control over the effect of corrective forces (e.g., forces arising from combining the wires' stiffness with the attachments' thickness) is never total control. Indeed, the transfer of mechanical energy in natural systems may result in dynamics opposing these forces or cooperating with them. It is necessary to visualise the critical points in the application to gain complete control over the outcome of the forces. The operator about to correct a crossbite must clarify the extent to which this condition constitutes an attractor.

From the logic of cause-effect explication (things take their truth from causes) to the logic of implication (the truth is the internal flow of processes): once again, in statistical regression procedures, causal factors are also caused, regressors are also regressed, the label is predicted but also predicts regressors. Because each patient has its cause and effect, computational systems can aid the reliability of prognostic predictions by highlighting underlying aspects of the facial system that are not clinically detectable; for instance, they can point to threshold values of parameters accompanied by unfavourable growth conditions. But that is not enough.

Since future evolution is perceived to be already at work in the present situation, converting into numerical terms the structure-event gap to feed a predictive machine cannot be a trivial operation. This operation necessitates reconfiguring the field of the thinkable in prognostic terms. In the system of craniofacial processes resulting from the interaction of multiple factors, the localisation of the 'tiny stage' of the clinical situation could rely on applying computational algorithms. A feature is a single measurable property of the phenomenon under observation. The orthodontist must search for the most discriminative, informative, independent, non-redundant feature. Regardless of the degree of complexity, every predictive system needs enhancement by data [12]. Raw data needs to become intelligent data for the period in which it is most meaningful and least polluted by noise.

The neutral zone of dynamic equilibrium toward which we strive to arrange the dental elements should not be a place of opposition of mutually exclusive forces but a place of meeting complementary forces that may or may not include and reinforce each other. Each pair is complementary, so each complement contains something of the other.

As orthodontists, we know that every phenomenon of existence possesses its opportune moment and legitimate place of fulfilment, both of which can be controlled and stabilised, at least partially, through the proper use of the alternation of the laws that govern them. Laws are not just about the tension-pressure relationship but also about hard wires, soft wires, extraoral traction-muscle forces, early treatment-waiting, and so on. It is necessary to shed light on what "is to happen" based on "what has just happened" because the one implies the other. In the ongoing unfolding, foresee in advance what will result and correct just enough to reorient and eventually stabilise through the propensity of the forces at play alone. The end of treatment is, at the same time, the beginning of a new transition. If we could perceive facial areas of stress accumulation, fulcrums of inertia, and vectors of distorting forces as quickly as we sense a local defect or entirely normal and healthy condition, this ability would become a substantial diagnostic and therapeutic tool.

From the moment it is triggered, a biological process unfolds. It begins something that only asks to become. We need not force the treatment effect and avoid saturation. The too-sharp point of a sword ends up breaking; too rigid a thread causes root ankylosis. We do not have to exceed the internal logic of the situation: the excess effect exceeds the tolerance threshold of the living matter, making it no longer integrable. The effect must not be the object of any overload. There is no need to overcorrect or add anything to the pure effect: reinforcing effect weakens the impact itself.

The ideal situation of living systems is "the steady state away from equilibrium": a background of virtuality should be operative when we can identify the distorting forces pre-existing the malocclusion we are correcting. We must verify the sustainability of overly symmetrised structures, enormously expanded arches, constrictive occlusal situations, dynamic stereotypies capable of triggering muscular hypertrophy, parasitic forces, and intrusive and distorting loads on a case-by-case basis. When at the end of treatment, we admire a pleasant smile, a harmonious occlusion, a perfect occlusal fit, let us ask ourselves whether it is possible to crystallise this situation over time, whether we have clarified the totality of the "boundary conditions."

Our facial features not only indicate who we are now but also encompass all that they have been and could become. Our destiny is to continuously change shape. Through maturation, we experience our potential and the fruits of what we have become. Computational systems function as automatic suggestors to remind us that our morphological reality has a greater extent than what appears. By keeping these aspects in mind, our diagnostic abilities can gain further perspectives, illuminating the areas where evolutionary possibilities coexist and the areas not yet manifested.

References

1. B. Liu, W. Hsu, Y. Ma, Pruning and summarizing the discovered associations, in *SIGKDD International Conference on Knowledge Discovery & Data Mining (KDD-99)*, 1999. Accessed 15 Dec 2023 [Online]. Available: https://www.comp.nus.edu.sg/~whsu/publication/1999/kdd99ds.pdf

2. Kaneman D. Thinking. fast and slow. New York: Farrar, Straus and Giroux, (2011).
3. M. Scutari, P. Auconi, G. Caldarelli, L. Franchi, Bayesian networks analysis of malocclusion data. Sci. Rep. **7**(1), 15236 (2017). https://doi.org/10.1038/s41598-017-15293-w
4. T. Wansbeek, E. Meijer, Measurement error and latent variables, in *A Companion to Theoretical Econometrics* (Wiley, 2003), pp. 162–179. https://doi.org/10.1002/9780470996249.ch9
5. F. Jullien, Conférence sur l'efficacité. 1951. Accessed: Oct. 07, 2023. [Online]. Available: https://www.puf.com/content/Conférence_sur_lefficacité_0
6. G. Caldarelli, *Scale-Free Networks: Complex Webs in Nature and Technology*, vol. 9780199211. (2007). https://doi.org/10.1093/acprof:oso/9780199211517.001.0001
7. M.E.J. Newman, The structure and function of complex networks. SIAM Rev. **45**(2), 167–256 (2003). https://doi.org/10.1137/S003614450342480
8. D.S. Falster, M. Westoby, Plant height and evolutionary games. Trends Ecol. Evol. **18**(7), 337–343 (2003). https://doi.org/10.1016/S0169-5347(03)00061-2
9. T. Smith, M. Huston, A theory of the spatial and temporal dynamics of plant communities. Vegetatio **83**(1–2), 49–69 (1989). https://doi.org/10.1007/BF00031680
10. D. Noble, A theory of biological relativity: no privileged level of causation. Interface Focus **2**(1), 55–64 (2012) https://doi.org/10.1098/rsfs.2011.0067
11. R. Ciancaglini, *Gnatologia e dolori oro-facciali : problemi e soluzioni* (Elsevier Masson, 2007). Accessed 29 Mar 2024 [Online]. Available: https://www.libreriauniverso.it/articolo/50886/9788821429224-gnatologia-e-dolori-orofacciali-problemi-soluzioni-ciancaglini-elsevier-masson
12. C.T. Butts, Revisiting the foundations of network analysis. Science **325**(5939), 414–416 (2009). (1979). https://doi.org/10.1126/science.1171022.

SPRINGER NATURE

GPSR Compliance

The European Union's (EU) General Product Safety Regulation (GPSR) is a set of rules that requires consumer products to be safe and our obligations to ensure this.

If you have any concerns about our products, you can contact us on ProductSafety@springernature.com

In case Publisher is established outside the EU, the EU authorized representative is:

Springer Nature Customer Service Center GmbH
Europaplatz 3
69115 Heidelberg, Germany

The manufacturer's authorised representative in the EU is Springer Nature Customer Service Centre GmbH, Europaplatz 3, 69115 Heidelberg, Germany. If you have any concerns regarding our products, please contact ProductSafety@springernature.com

Printed and bound by CPI Group (UK) Ltd, Croydon, CR0 4YY

01/10/2026

02225916-0015